Efficiency Measures in the Agricultural Sector

Armando B. Mendes
Emiliana L.D.G. Soares da Silva
Jorge M. Azevedo Santos
Editors

Efficiency Measures in the Agricultural Sector

With Applications

Springer

Editors
Armando B. Mendes
Professor of Statistics and Operational Research
Researcher of Centre of Applied Economics Studies of the Atlantic (CEEAplA)
University of Azores
Portugal

Emiliana L.D.G. Soares da Silva
Professor of Agrarian Economy
Researcher of Centre of Applied Economics Studies of the Atlantic (CEEAplA)
University of Azores
Portugal

Jorge M. Azevedo Santos
Professor of Mathematics
School of Sciences and Technology
Researcher of Centre for Research in Mathematics and Applications (CIMA)
University of Évora
Portugal

ISBN 978-94-007-5738-7 ISBN 978-94-007-5739-4 (eBook)
DOI 10.1007/978-94-007-5739-4
Springer Dordrecht Heidelberg New York London

Library of Congress Control Number: 2012955153

Printed on acid-free paper

Springer is part of Springer Science+Business Media (www.springer.com)

Relevant Disciplines

Economics/Management Science – Agricultural Economics
Statistics – Statistics for Business/Economics
Mathematics – Optimization

Keywords

Data Envelopment Analysis
Stochastic Frontier Analysis
Agricultural Sector
Farm efficiency
Decision Support

Acknowledgments

This book was only possible with the dedication, interest and contribution of persons and institutions. We start by thanking the authors who replied to our challenge. Their work is based in researches/papers on agricultural efficiency using different methodologies in different contexts. This project was jointly organized by the Centre of Applied Economics Studies of the Atlantic (CEEAplA) of University of Azores and the Centre for Research in Mathematics and Applications (CIMA) of the University of Évora. We also express our appreciation to the referee committee, Professors Ana Alexandra Marta-Costa (UTAD, CETRAD), Francisco Silva (University of Azores, CEEAPLA), José Cabral Vieira (University of Azores, CEEAPLA) and Luís Cavique (Open University, LabMAg). Finally, we would like to thank Dr. Susana Sardinha and Dr. Sérgio Frias for editing the English language and all thc formatting in this book and Springer for embracing this project. A special thanks is due to Professor Veska Noncheva for the original idea of the project.

Armando B. Mendes
Emiliana Silva
Jorge Santos

This work has been supported by Regional Division for Science and Technology of Azores Government through the project M.2.1.2/l/009/2008, "Productivity Analysis of Azorean Cattle-Breeding Farms with R Statistical Software".

Contents

Contributors

Amílcar Arzubi Lomas de Zamora University, Lomas de Zamora, Argentina

Julio Berbel ETSIAM, Department of Agricultural Economics, Córdoba University, Córdoba, Spain

Rafaela Dios-Palomares University of Córdoba, Córdoba, Spain

Magdalena Kapelko Institute of Applied Mathematics, Wroclaw University of Economics, Wroclaw, Poland

Alfons Oude Lansink Business Economics Group, Wageningen University, Wageningen, The Netherlands

Eusébio Marote CEEAplA, Departamento de Ciências Agrárias, Universidade dos Açores, Angra do Heroísmo, Açores, Portugal

José M. Martínez-Paz University of Murcia, Campus Espinardo, Murcia, Spain

Armando B. Mendes CEEAplA, Departamento de Matemática, Universidade dos Açores, Ponta Delgada, Açores, Portugal

Elsa Rosário Negas CITAD, Lusíada University of Lisbon, Portugal

Veska Noncheva CEEAplA, Faculty of Mathematics and Informatics, University of Plovdiv, "PaisiiHilendarski" Plovdiv, Bulgaria

Angel Prieto IRNASA. CSIC, Salamanca, Spain

Ana Sampaio ISCTE Business School Scientific Research & Development Unit, Évora University, Évora, Portugal

Carlos Santos Departamento de Ciências Agrárias, Universidade dos Açores, Angra do Heroísmo, Açores, Portugal

Jorge M. Azevedo Santos CIMA, Departamento de Matemática, Universidade de Évora, Évora, Portugal

Luís Cavique Santos Science and Technology Department, Universidade Aberta, Lisbon, Portugal

Emiliana L.D.G. Soares da Silva CEEAplA, Departamento de Ciências Agrárias, Universidade dos Açores, Angra do Heroísmo, Açores, Portugal

Fátima Venâncio CEEAplA, Departamento de Ciências Agrárias, Universidade dos Açores, Angra do Heroísmo, Açores, Portugal

List of Figures

List of Tables

Part I
Efficiency Measures and Methods

Chapter 1
Efficiency Measures in the Agricultural Sector: The Beginning

Emiliana Silva, Armando B. Mendes, and Jorge Santos

Abstract The agricultural productivity is often based on non-parametric models (DEA), or stochastic models (SFA). In this initial article, the editors start by pointing that the models (DEA and SFA) allow estimating the efficiency of the production frontier and their structural forms. Then, it is presented, in general terms, the differences between DEA and SFA models: DEA model involves the use of technical linear programming to construct a non-parametric piecewise surface, and SFA models comprise econometric models with a random variable, or an error term, including two components: one to account for random effects and another to take care of the technical inefficiency effects. Finally, it shows a comparison between the two approaches (SFA and DEA) and the advantages and disadvantages of their utilizations.

Keywords Data Envelopment Analysis • Models • Non-parametric • Parametric • Stochastic Frontier Analysis

E. Silva (✉)
CEEAplA, Departamento de Ciências Agrárias, Universidade dos Açores, Rua Capitão João de Ávila, 9700-042 Angra do Heroísmo, Açores, Portugal
e-mail: emiliana@uac.pt

A.B. Mendes
CEEAplA, Departamento de Matemática, Universidade dos Açores, Rua da Mãe de Deus, 9500-801 Ponta Delgada, Açores, Portugal
e-mail: amendes@uac.pt

J. Santos
CIMA, Departamento de Matemática, Universidade de Évora, Colégio Luís Antº Verney R. Romão Ramalho, 59 7000-671 Évora, Portugal
e-mail: jmas@uevora.pt

A.B. Mendes et al. (eds.), *Efficiency Measures in the Agricultural Sector: With Applications*, DOI 10.1007/978-94-007-5739-4_1,

1.1 Introduction

The agricultural policy analysis, mainly, the agricultural productivity, is often based on support models like mathematical programming (non-parametric, non-stochastic) models or econometric (stochastic, parametric) models. These models are very useful for decision support because they give an orientation of what are the main characteristics of agricultural farms and how it is possible to solve some of the problems found.

The deterministic production frontier is an approach where all observations are in one side of the frontier and all deviations from the frontier are attributed to inefficiency. On the other hand, in the stochastic approach, all observations are on both sides of the frontier, and it is possible to separate between random errors and differences in inefficiency.

The most popular approaches to calculate efficiency are: (1) the non-parametric techniques (Charnes et al. 1978), the Data Envelopment Analysis (DEA) based on the linear programming tools; and (2) parametric techniques (Aigner et al. 1977; Meeusen and van den Broeck 1997), the Stochastic Frontier Analysis (SFA) – stochastic frontier production (SFP) – based on econometric tools.

1.2 The Data Envelopment Analysis: DEA

The use of optimization tools to calculate efficiency with Data Envelopment Analysis (DEA) was developed by Charnes et al. (1978) from earlier work by Farrell (1957). This method has been used to estimate the efficiency in the organizational units in several areas (Cooper 1999).

DEA is a non-parametric method to estimate efficiency, involves the use of technical linear programming to construct a non-parametric piecewise surface (or frontier) over data, for it to be able to calculate efficiency relative to his surface (Coelli 1996a). Any farm that lies below the frontier is considered to be inefficient. DEA permits to construct a best-practice benchmark from the data on inputs and outputs (Jaforullah and Whiteman 1999). In opposite, parametric techniques, as econometric methods, construct a stochastic frontier.

DEA involves the concept of efficiency, and Farrell (1957) had decomposed the efficiency into (1) technical efficiency and (2) allocative efficiency. The technical efficiency measures the maximum equiproportional reduction in all inputs, which still allows continued production of given outputs. The allocative efficiency reflects the ability of a firm to use the inputs in optimal proportions, given their respective prices. These two concepts form the concept of economics efficiency (Coelli 1995). The allocative inefficiency measures the magnitude of consequent loss. Similar considerations are applied to economics efficiency and inefficiency.

Therefore, the overall measure of technical efficiency can be disaggregated into three components: (1) pure technical efficiency due to producing within an isoquant frontier, (2) congestion due to overutilization of inputs and (3) scale efficiency due to deviations from constant returns to scale (Weersink et al. 1990).

Technical efficiency under constant returns to scale is estimated by relating each observation to the frontier under constant returns to scale (CRS). Technical efficiency under variable returns to scale (TEVRS) and technical efficiency under constant returns to scale (TECRS) are equal for farms that operate in the region of constant returns to scale, i.e. these farms have a scale efficiency of one. As a consequence, these farms have a scale efficiency (SCAL) of one (Lansink and Reinhard 2004).

One of the most popular computer programs used to solve DEA problems is the Data Envelopment Analysis Computer Program (DEAP) developed by Coelli (1996b). This program is based on the optimization model used by Charnes et al. (1978), but considering the input components $v_{ik}x_{ik}$ and a scaling constant l (normally 100%):

$$\text{Max}: \; Ef_a = \sum_{r=1}^{s} \mu_{ra} y_{ra}$$

$$\text{s.t.}: \sum_{i=1}^{m} v_{ia} x_{ia} = l$$

$$\sum_{r=1}^{s} \mu_{rk} y_{rk} \leq \sum_{i=1}^{m} v_{ik} x_{ik} \quad k = 1, \ldots, n$$

$$\mu_{rk}, v_{ik} \geq 0 \quad i = 1, \ldots, m \quad r = 1, \ldots, s \tag{1.1}$$

y_{rk} is the level of output r used by decision-making unit k, x_{ik} is the level of input i used by decision-making unit k and μ_{rk} and v_{ik} are the non-negative variable weights associated to the solution of decision-making unit k, of output (r) and inputs (i), respectively; s is the number of outputs considered and m the number of inputs considered.

The DEA approach has been applied to the agricultural field to estimate the efficiency by different researchers in different parts of the world, such as Arzubi and Berbel (2002), Reinhard and Thijssen (2000), Reinhard et al. (2000), Jaforullah and Whiteman (1999), Fraser and Cordina (1999), Gonzalez et al. (1996), Färe and Whittaker (1995) and Weersink et al. (1990).

1.3 Stochastic Frontier Analysis (SFA) or Production (SFP)

The Stochastic Frontier Analysis (SFA) is a parametric approach which was originally and independently proposed by Aigner et al. (1977) and Meeusen and van den Broeck (1977) as pointed out by Battese and Coelli (1988). The SFA involves an unobservable random variable associated with the technical inefficiency in production of individual firms. In addition to the random error in a traditional regression model (Battese and Broca 1997), the SFA considers that an error term has two components: one to account for random effects and another to account for the technical inefficiency effects (Coelli 1995).

The most common different functional forms for stochastic frontier are the tanslogarithmic and the Cobb-Douglas production functions.

One of the most common software to solve parametric problems is FRONTIER version 4.1, developed by Coelli (1996b), often selected to estimate the models.

The efficiency model (*model I*) is proposed by Battese and Coelli (1995), and the v_{it} is the technical inefficiency variable which is defined by

$$v_{it} = \{\exp[-\eta(t-T)]\}\, v_i, \quad \text{with} \quad i = 1, \ldots, N, \quad \text{and} \quad t = 1, \ldots, T \qquad (1.2)$$

where η is an unknown parameter to be estimated; and v_i with $i = 1, 2, \ldots, n$, are independent and identically distributed non-negative random variables, which are assumed to account for technical inefficiency in production and obtained by truncations (at zero) of the normal distribution, with unknown mean and unknown variance, σ^2.

This model specifies that the technical inefficiency effects for the sample in the earlier periods of the panel are a deterministic exponential function of the inefficiency effects for the corresponding firms in the last years of the panel (i.e. $v_{it} = v_i$, given the data for the *i-th* firm is available in the period, T).

Given the Eq. (1.2), the expectation of the mean technical efficiency is

$$\text{TE} = \exp(-v_i), \qquad (1.3)$$

and it is estimated considering the technical inefficiency effects (v_i).

The *model I* only permits to determine the technical efficiency, but to know more about inefficiency, Battese and Coelli (1995) presented the *model II* (inefficiency model) which permits to incorporate in *model I* the variable that could cause inefficiency in the firms. In this model, the technical inefficiency effects are defined by

$$u_{it} = z_{it}\delta + w_{it}, \quad \text{with} \quad i = 1, \ldots, N, \quad \text{and} \quad t = 1, \ldots, T \qquad (1.4)$$

where z_{it} is a $(1 \times M)$ vector of explanatory variables associated with technical inefficiency effects; δ is a $(M \times 1)$ vector of unknown parameters to be estimated; and w_{it} are the unobservable random errors which are assumed to be independently distributed and obtained by truncation of a normal distribution with unknown mean and variance, σ^2, such that u_{it} is non-negative ($w_{it} \geq -z_{it}\delta$).

This approach is widely used by many authors such as Pascoe et al. (2001), Torres et al. (2002), Munzir and Heidhues (2002), Reinhard et al. (1999), Reinhard et al.(2000), Alvarez and Gonzalez (1999), Webster et al. (1998), Franco (2001), Daryanto et al. (2002), Lawson et al.(2004), Battese and Coelli (1988), Battese and Broca (1997), Battese and Coelli (1992), Brummer (2001), Hallam and Machado (1996) and Venâncio and Silva (2004).

1.4 DEA and SFA Approaches: A Comparison

DEA is very useful to calculate efficiency (Lansink and Reinhard 2004) and provides a simple way to calculate the efficiency gap that separates each producer's behaviour from the best practice, which can be assessed from actual observations of the inputs and outputs of efficient firms (Reig-Martínez and Picazo-Tadeo 2004).

The DEA model allows the comparison of a firm to a benchmark (set of best producers), and then the measure of efficiency is relative to the best producer in that group of firms; it is not necessarily the maximum real output per input used.

Sharma et al. (1999) argue that the main advantage of DEA is the fact that avoids the parametric specification of technology such as the assumptions for the distribution of the inefficiency term.

The DEA approach has the advantage of considering many inputs and many outputs simultaneously. This has as a consequence the increasing of efficiency with the number of variables: more variables, higher efficiency (Reig-Martínez and Picazo-Tadeo 2004; Silva et al. 2004). DEA does not require a parametric specification of a functional form to construct the frontier (Silva et al. 2004), and it was considered by Coelli and Perelman (1999) as the main advantage of DEA; it is very easy to perform because it does not require a priori knowledge on the functional form of the frontier and benchmarking (best-practice reference) to real firms (Lauwers and van Huylenbroecks 2003). DEA does not impose any assumptions about functional form; hence, it is less prone to misspecification and does not take into account the random error and consequently is not subject to the problems of assuming an underlying distribution about error term (Pascoe and Mardle 2000). To these authors, DEA does not take account of statistical noise, and the efficiency estimates may be biased if the production process is largely characterized by stochastic elements.

The mathematical programming models are non-stochastic, and so they cannot have the values, of inefficiency and noise, separately. It is a non-parametric technique and not as effective as SFA, as a specification error (Reig-Martínez and Picazo-Tadeo 2004). The major limitations of DEA is that it is difficult, conceptually, to separate the effects of uncontrollable environmental variables and measurements error, from the effect of differences in farm management and the presence of outliers (Silva et al. 2004).

The DEA model is deterministic and attributes all the deviations from the frontier to inefficiencies; a frontier estimated by DEA is likely to be sensitive to measurement errors or other noise in data (Sharma et al. 1999) and may attribute stochastic noise to the inefficiency scores and thus may be more sensitive to outliers (Lauwers and van Huylenbroecks 2003).

DEA has more flexibility in that they avoid a parametric specification of technology and assumptions about the distribution of efficiency, whilst allowing curvature conditions to be imposed easily (Lansink and Reinhard 2004).

The main DEA disadvantage to Coelli and Perelman (1999) is that when the calculation of shadow prices are desired, only a range of prices can be derived for the efficient firms. The production surface constructed by DEA is a series of intersecting planes. The efficient frontier points that define this frontier surface (primarily) lie at the intersections of these planes. Hence, when one attempts to measure shadow prices for these efficient points, only a range of price ratios can be observed (corresponding to the slopes of the planes involved).

A disadvantage of the DEA approach is that there are no single objective criteria (it is different from a CRS to a VRS) against assessment of the model, and the models only provide a reasonable representation of the actual frontier (or set of frontiers) (Pascoe and Mardle 2000).

The econometric approach is stochastic, and then it allows distinguishing the two effects: statistical noise from productive inefficiency. It is also parametric and can mix the effect of misspecification of functional form (can be flexible) with inefficiency. The flexible functional form could imply the multicollinearity, and some theoretical conditions could be violated (Reinhard et al. 2000)

One of the most important characteristics of econometric models (SFA) is that it allows a specification in the case of panel data and the construction of confidence intervals (Reinhard et al. 2000).

The SFA models only have in this functional form one output and various inputs. SFA allows a correction for stochastic events, but assumes a parametric specification for the production technology, which can confuse efficiency results. SFA calculated by a translog specification, the curvature conditions (concavity inputs) are not globally satisfied, SFA makes an explicit assumption about the distribution of the inefficiency term (Lansink and Reinhard 2004).

Stochastic estimations incorporate a measure of random error but impose an explicit functional form and distribution assumption of data (Pascoe and Mardle 2000). SFP approach produces a set of statistics against which the models can be judged in terms of goodness of fit and alternative methods can be discriminated against.

One advantage of parametric methods is that they permit the testing of hypotheses such as those relating to the significance of included inputs and/or outputs, returns to scale and so on (Coelli and Perelman 1999).

The advantage of DEA over SFA is that the technological frontier is constructed without imposing a parametric functional form for inefficiency or on a deviation from it (Reig-Martínez and Picazo-Tadeo 2004).

Sharma et al. (1999) found that DEA is more sensitive to outliers and other noise in the data, but DEA results are more robust than those obtained from the parametric approach.

The most important point of frontier approach is that it deals with stochastic noise and allows statistical test of hypotheses pertaining to production structure and the degree of inefficiency. The imposed structural form and the explicit distributional assumption for the inefficiency term are the worst point of parametric techniques (Sharma et al. 1999).

There are several differences between these two kinds of tools. One of the differences pointed by Reinhard et al. (2000) was about the construction of a production frontier and the calculation of efficiency relative to the frontier.

Coelli and Perelman (1999) point as advantages of parametric and non-parametric approaches the influence of outliers, although the stochastic frontier method attempts to account this problem, the translog distance function estimates obtained by this tool, has not been very successful.

In some situations in DEA, when there are non-stochastic elements affecting a firm's level of production, the greater DEA flexibility allows a better representation in the level of the distribution of technical efficiency (Pascoe and Mardle 2003). In this case, the imposition of a single production frontier (SFA model) may result that some inefficient firms probably are efficient.

Paul et al. (2004) compare the stochastic and determinist methods and conclude that in terms of level, the SFP (Stochastic Frontier Production) measures suggest more scale economies and technical efficiency, and DEA measures more "technical progress" or shifts out of the technological frontier over time.

In the construction of confidence intervals, Brummer (2001) points that intervals for SFA are wider than DEA, because the assumptions of DEA are more restrictive (deterministic structure). The SFP model estimated a higher mean technical efficiency compared with DEA models (Pascoe and Mardle 2000).

Reinhard et al. (2000) compared the SFA and DEA approaches and concluded that both methods can estimate efficiency scores, but only SFA incorporates noise. DEA is a deterministic model and could not identify the environmental detrimental variables in the model.

In summary, the main difference between SFA and DEA is about the way the production possibility is estimated; that means, DEA does not require a functional form and SFA does. The DEA (deterministic) approach ignores the random effects (noise), but SFA approach takes it into account; then in the DEA model, any deviation is considered inefficiency but in SFA is considered noise and inefficiency. Both approaches give the same indication of the characteristics that affect efficiency, and the outliers can affect the scores of efficiency. Although the scores of inefficiency are similar, usually the value of the DEA approach is lower than the one obtained by SFA. The SFA approach allows distinguishing the inefficiency from the error term: it is identified what is due to inefficiency and what is due to random disturbances such as measurement of error, luck, bad weather, pest or diseases.

Both approaches seem useful, and their use will depend on the objectives of the analysis.

1.5 Why This Book?

The aim of this book is to aggregate several articles on efficiency measures and techniques for the agricultural sector.

Increasing efficiency in the agricultural sector and in rural communities can improve the financial situation of farms and communities, given the continuous pressure on margins of agricultural products. This fact has been highly recognized and studied by researchers since many years. But, in the current context of economic crises in Europe, these themes are, more than ever, especially pertinent as several southern countries need to increase exportation and decrease importation.

This book starts by describing techniques like Data Envelopment Analysis (DEA) and Stochastic Frontier Analysis (SFA) and proceeds by presenting several applications exploring the particularities of the agriculture sector as the abundance of subsidies, strong competition between regions worldwide and the need for efficient production systems. The introductory chapters make possible for the nonexpert reader to understand the book, and clear applications lead the reader to practical problems and solutions.

Applications include the estimation of technical efficiency in agricultural grazing systems (dairy, beef and mixed) and specifically for dairy farms. The conclusions indicate that it is now necessary to help small dairy farms in order to make them more efficient. These results can be compared with the technical efficiency of a sample of Spanish dairy processing firms presented by Magdalena Kapelko and co-authors.

In the work by Silva and Venâncio, SFA is used for the estimation of inefficiency models. Another approach involved the assessing of the importance of subsidies in farms efficiency as they are so relevant for farmers. Another application studies the level of technical efficiency in the Andalusia oil industry implementing environmental and non-discretionary variables.

There are also articles on applying DEA techniques to agriculture using R software and making it more user friendly. The "Productivity Analysis with R" (PAR) framework establishes a user-friendly data envelopment analysis environment with special emphasis on variable selection, aggregation and summarization.

The editors have confidence that this book is useful and informative for students and researchers. Feel free to send feedback to amendes@uac.pt.

References

Aigner D, Lovell C, Schmidt P (1977) Formulation and estimation of stochastic frontier production functions models. J Econ 6:21–37

Alvarez A, Gonzalez E (1999) Using cross-section data to adjust technical efficiency indexes estimated with panel data. Am J Econ 81:894–901

Arzubi A, Berbel J (2002) Determinación de índices de eficiencia mediante DEA en explotaciones lecheras de Buenos Aires. Invest Ag Prod Sanid Anim 17(1–2):103–123

Battese G, Broca S (1997) Functional forms of stochastic frontier production function and models for technical inefficiency effects: a comparative study for wheat farmers in Pakistan. J Product Anal 8:395–414

Battese G, Coelli T (1988) Prediction of firm-level technical efficiencies with a generalized frontier production function and panel data. J Econ 38:387–399

Battese G, Coelli T (1992) Frontier production functions. Technical efficiency and panel data: with application to paddy farmers in India. J Product Anal 3:153–169

Battese G, Coelli T (1995) A model for technical inefficiency effects in a stochastic frontier production function for panel data. Empir Econ 20:325–332

Brummer B (2001) Estimating confidence intervals for technical efficiency: the case of private farms in Slovenia. Eur Rev Agric Econ 28(3):285–306

Charnes A, Cooper W, Rhodes E (1978) Measurement the efficiency of decision making units. Eur J Oper Res 2:429–444

Coelli T (1995) Recent developments in frontier modeling and efficiency measurement. Aust J Agric Econ 39:219–245

Coelli T (1996a) A guide to DEAP Version 2.1. A data envelopment analysis computer program. Centre for efficiency and productivity analysis. Department of Econometrics, University of New England, Armidale, Australia

Coelli T (1996b) A Guide to FRONTIER Version 4.1: a computer program for stochastic frontier production and cost function estimation, CEPA Working paper no. 96/07. Department of Econometrics, University of New England, Armidale

Coelli T, Perelman S (1999) A comparison of parametric and non-parametric distance functions: with application to European railways. Eur J Oper Res 117:326–339

Cooper WW (1999) Operational research/management science – where it's been – where it should be going. J Oper Res Soc 50:3–11

Daryanto H, Battese G, Fleming E (2002) Technical efficiencies of rice farmers under different irrigation systems and cropping seasons in West Java, University of New England, Asia Conference on Efficiency and Productivity Growth. Retrieved April 20, 2003, from http://www.sinica.edu.tw/~teps/A1-1.pdf

Färe R, Whittaker G (1995) An intermediate input model of dairy production using complex survey data. J Agric Econ 46(2):201–213

Farrell M (1957) The measurement of productive efficiency. J R Stat Soc (Ser A) 120(III):253–290

Franco F (2001) Eficiência Comparada Dos Serviços Hospitalares: O Método de Fronteira Estocástica, Tese de Mestrado, Universidade dos Açores

Fraser I, Cordina D (1999) An application of data envelopment analysis to irrigated dairy farms in Northern Victoria, Australia. Agric Syst 59:267–282

González E, Álvarez A, Arias C (1996) Análisis no paramétrico de eficiencia en explotaciones lecheras, Investigación Agraria: Economía 11(1):173–190

Hallam D, Machado F (1996) Efficiency analysis with panel data – a study of Portuguese dairy farms. Eur Rev Agric Resour Econ 23(1):79–93

Jaforullah M, Whiteman J (1999) Scale efficiency in the New Zealand dairy industry: a non-parametric approach. Aust J Agric Resour Econ 43(4):523–541

Lansink A, Reinhard S (2004) Investigating technical efficiency and potential technological change in Dutch pig farming. Agric Syst 79:353–367

Lauwers L, van Huylenbroecks G (2003) Materials balance based modeling of environmental efficiency. Presented at the 25th international conference of agricultural economist, South Africa, 16–22 Aug 2003

Lawson L, Bruun J, Coelli T, Agger J, Lund M (2004) Relationships of efficiency to reproductive disorders in Danish milk production: a stochastic frontier analysis. J Dairy Sci 87:212–224

Meeusen W, van den Broeck J (1977) Efficiency estimation from Cobb-Douglas production function with composed error. Int Econ Rev 18:435–444

Munzir A, Heidhues F (2002) Towards a technically efficient production in rural aquaculture, case study at Lake Maninjau. Presented at Indonesia, International Symposium Sustaining Food Security and Managing Natural Resources in Southeast Asia – Challenges for the 21st Century, Chiang Mai, Thailand. Retrieved July 21, 2003 from http://www.uni-hohenheim.de/symposium2002/pa_full/Full-Pap-S3B-3_Munzir.pdf

Pascoe S, Mardle S (2000) Technical efficiency in EU fisheries: implications for monitoring and management through efforts controls: "TEMEC". EU funded project QLK5-CT1999-01295. University of Portsmouth, pp 1–5

Pascoe S, Mardle S (2003) Single output measures of technical efficiency in EU fisheries. CEMARE Report 61, CEMARE, University of Portsmouth, UK

Pascoe S, Hassaszahed P, Anderson J, Korsbrekke K (2001) Economic versus physical input measures in the analysis of technical efficiency in fisheries. Presented at the XII Conference of the European Association of Fisheries Economists, Italia. Retrieved April 25, 2003 from http://www.eafe-fish.org/conferences/salerno/papers/paper04_seanpaschoe.doc

Paul C, Nehring R, Banker D, Somwaru A (2004) Scale economies and efficiency in U.S. agricultures: are traditional farms history? J Product Anal 22:185–205

Reig-Martínez E, Picazo-Tadeo A (2004) Analysing framing systems with data envelopment analysis: citrus farming in Spain. Agric Syst 82(1):17–30

Reinhard S, Thijssen G (2000) Nitrogen efficiency of Dutch dairy farms: a shadow cost system approach. Eur Rev Agric Econ 27(2):167–186

Reinhard S, Lovell K, Thijssen G (1999) Econometric estimation of technical and environmental efficiency: an application to Dutch dairy farm. Am J Agric Econ 81:44–60

Reinhard S, Lovell K, Thijssen G (2000) Environmental efficiency with multiple environmentally detrimental variable: estimated SFA e DEA. Eur J Oper Res 121:287–303

Sharma K, Leung P, Zaleski H (1999) Technical, allocative and economic efficiencies in swine production in Hawaii: a comparison of parametric and nonparametric approaches. Agric Econ 20:23–35

Silva E, Arzubi A, Berbel J (2004) An application of data envelopment analysis (DEA) in Azores dairy farms, Portugal. New MEDIT III(3):39–43

Torres J, Basch M, Vergara S (2002) Eficiência Técnica Y Escalas de Operación en Pesca Pelágica: Un Análisis de fronteras estocásticas, Universidade do Chile. Retrieved April 21, 2003 http://www.ilades.cl/economia/Publicaciones/ser_inv/inv137.pdf

Venâncio F, Silva E (2004) A Eficiência de Explorações Agro-pecuárias dos Açores: uma abordagem paramétrica, Actas XIV Jornadas Luso – Espanholas de Gestão Científica

Webster R, Kennedy S, Johnson, L. (1998). Comparing techniques for measuring the efficiency and productivity of Australian private hospitals. Working Papers (98/3) in Econometrics and Applied Statistics, Australian Bureau of Statistics. Retrieved May 9, 2003 from http://www.abs.gov.au/websitedbs/D3110122.NSF/0/31f3a2cdb3dbbf85ca25671b001f1910/$FILE/13510_Nov98.pd

Weersink A, Turvey C, Godah A (1990) Decomposition measures of technical efficiency for Ontario, dairy farms. Can J Agric Econ 38:439–456

Chapter 2
Review of Frontier Models and Efficiency Analysis: A Parametric Approach

Ana Sampaio

Abstract The parametric frontier approach to efficiency measurement has been extensively used in applied research. Within this conceptual framework, techniques for econometric frontier analysis will be described. The purpose of this paper is to present an overview of parametric frontier methods related to the measurement of economic efficiency, focusing on both deterministic and stochastic perspectives. In addition, development and extension of the cross-sectional and panel data context associated with specification of functional forms are also revisited.

Keywords Efficiency analysis • Parametric frontier models • Functional forms • Stochastic and deterministic specification

2.1 Introduction

In the actual context of global economy, with some economies experiencing slow and decelerating growth, accompanied by high levels of unemployment, sustainable economic recovery of states emerges as a priority issue of world development strategy. It is in this paradigm that organizations' competitiveness, allied to efficiency analysis, must be allocated as tools to improve societies well-being. Also at the micro-level of analysis, efficiency is associated to sustainable development as the concept evolves the parsimony use of economic resources in order to reach cost minimization, output and profit maximization. Measurement of firm efficiency represents one of the most important subjects of investigation at the microeconomic level, either in the context of developing and developed countries or within different contexts of analysis. This is supported by the amount of empirical studies dealing

A. Sampaio (✉)
ISCTE Business School Scientific Research & Development Unit, Évora University,
Rua Romão Ramalho, 59, 7000-671 Évora, Portugal
e-mail: sampaio@uevora.pt

A.B. Mendes et al. (eds.), *Efficiency Measures in the Agricultural Sector: With Applications*, DOI 10.1007/978-94-007-5739-4_2,

with analysis of organizational efficiency emerging over the past 30 years in benchmarking scientific literature, covering a wide range of fields. Agriculture, banking, environmental economics, finance, transport, education, forestry, tourism and sport are examples of sectors where such evaluation has been applied. The research of reasons behind firm inefficiency is very important as it may be used to correct fragilities and to improve efficiency in the production context. According to the literature, organization' efficiency measures may be obtained through the estimation of an efficient frontier, with distance to this frontier being an indicator of the organization inefficiency (Kopp and Diewert 1982). Although traditional regression analysis has been widely applied in economic analysis, for the study of firm efficiency, it is consensual among the scientific community that frontier methodology is much more appropriate than least-squares methods, because the adjustment of a function through the middle of a cloud of points can only obtain average parameter estimates of the production structure rather than those associated with 'best practice' technology (Greene 1997). The concept of (in)efficiency is associated with the theory of optimization and with the extent to which an economic unit fails to achieve a theoretically ideal level of production possibilities (Forsund et al. 1980).

The research of reasons behind firm inefficiency is very important as it may be used to correct fragilities and to improve efficiency in the production context. Efficiency is associated with (1) technical efficiency if the goal of the analysis is to obtain maximum output given a set of inputs, (2) cost-efficiency if the aim is the minimum cost of producing that output given the input prices or (3) profit efficiency in the case where interest is in the maximum profit attainable given the inputs, outputs and price of the inputs (Greene 2005).

In order to measure organizations' economic (in)efficiency, two main alternative approaches have been developed and applied in empirical studies: a parametric and a non-parametric method, or stochastic frontier approach (SFA) and data envelopment analysis (DEA) (Charnes et al. 1978), as the most efficient frontier is best described by a parametric function involving econometric methods for estimation purposes or constructed through the use of a mathematical programming model applied to observed data, respectively.

Frontier models and the associated measurement of efficiency have a long history in the field of applied economics. Indeed and although papers by Debreu (1951) and Koopmans (1951) had marked the origin of discussion on measurement of efficiency, it was the work of Farrell (1957), extending Debreu and Koopmans researches that suggested to measure technical (in)efficiency as the realized deviation from a frontier isoquant. After Farrell's seminal article on efficiency measurements, several other approaches have been developed. It is consensual that only after the 1950s was the threshold between micro- and macro-level production/cost analysis developed, expanding economic analysis at firm level.

For contextualization of the parametric approach to economic frontier modelling at the microeconomic level, the contributions, in estimating production functions, of Cobb and Douglas (1928), Samuelson (1938), Dean (1951), Shephard (1953), Johnston (1959), Arrow et al. (1961) and Nerlove (1963) represent a significant point in econometrical literature. Still in the paradigm of empirical analysis at the micro context, extension and identification of the term inefficiency, with the usual disturbance component of a regression model, have been frequently applied.

This chapter presents an overview of the parametric frontier approach to efficiency measurement. Section 2.2 refers to the econometric approach to efficiency, Sect. 2.3 reviews cross-sectional frontier models, Sect. 2.4 deals with frontier analysis with panel data and, finally, Sect. 2.5 extends the theory to new statistical developments and applications.

2.2 Econometric Approach to Efficiency

The econometric approach to efficiency results from the estimation of frontier models which deal with parametric representations of technology along with a one-sided error term or a two-part error term (composed error), depending on how the functional form has been specified as being a deterministic or a stochastic frontier (Kalirajan and Shand 1999; Murillo-Zamoran and Vega-Cervera 2001). As the parametric approach to efficiency states an econometric specification, models may also be classified according to the function form, production, cost or profit, or functional form, which describes the relationship between inputs and outputs. Additionally, other criteria may be assumed to classify frontier models, such as the sample context (cross-sectional or panel data), the temporal variation of inefficiency and the functional specification of the inefficiency term of the composed error.

Introduced by Koopmans (1951), extended by Debreu (1951) and developed in the empirical econometric field through the seminal paper by Farrell (1957), the concept of economic efficiency has been of interest to modern researchers. Both (1) the conceptual contribution of Koopmans (1951), defining technical efficiency as a feasible input–output combination where it is not possible to increase output (decrease input) without simultaneously increasing input (decreasing output), and (2) the Debreu coefficient of resource cost allocation for the measurement of technical and allocative inefficiency, supported by resources, technology and preferences and resulting from the ratio between minimized resource costs of obtaining a given consumption bundle and actual costs, for given prices and a proportional contraction of resources (Forsund and Sarafoglou 2002), represent significant contributions to interpretation of efficiency. Augmenting the Debreu coefficient of a proportionate input contraction and associating technical inefficiency with deviations from an idealized frontier isoquant, Farrell' work has been greatly analysed and discussed in the context of empirical literature on production and cost functions. Extending the standard definition of production function, a production frontier model defines the boundary of the former as it gives the maximum possible output for a given set of inputs. So, an efficient organization will be located on the frontier of production, reflecting technical or allocative efficiency. Technical efficiency occurs when a firm produces at the maximum level of output given inputs or as it uses the minimum level of inputs given output. Allocative efficiency occurs when the marginal rate of substitution between inputs equals the input price ratio, or as it operates at an optimal proportion use of inputs, given the prices.

2.3 Cross-Sectional Frontier Models

In a cross-sectional context and depending on the specification of the parametric frontier function error term (one or two component), models used for the estimation of technical efficiency can be classified as deterministic frontier models or as stochastic frontier models. The parametric approach to efficiency measurement estimation appears in the production theory with Aigner and Chu (1968) being the first to follow Farrell's suggestion, but it is in the 1970s that the study of economic efficiency in the sectional context develops most. In the econometric literature, various criteria exist for classifying frontier models (Forsund et al. 1980).

The models are statistical when the error term is specified by a given distribution of probability and the estimators have statistical properties. On the contrary, models are not statistical when this term does not have statistical properties. In nonstatistical deterministic models, linear programming and quadratic programming techniques are used to construct the frontier.

In the category of statistical models, they are deterministic or stochastic according to the error term of the respective objective function being formed by one component, which only translates inefficiency of the process or by two components, an inefficiency term and a white noise, respectively. Parametric deterministic models evolve both goal programming and econometric techniques in order to either calculate the parameter vector or estimates of inefficiency.

With goal programming techniques, the technology parameter vector is calculated through the solution of a deterministic optimization method (Aigner and Chu 1968; Timmer 1971; Forsund and Hjalmarsson 1979; Nishimizu and Page 1982; Forsund 1992). This type of frontier was subsequently abandoned and replaced by another type more suited to statistical analysis of the results, that is, by deterministic parametric frontiers with statistical properties.

With the deterministic statistical approach, the parameters are estimated rather than calculated, allowing the use of additional statistical inference procedures. This new type of model, initially developed by Afriat (1972), was later enhanced by the contributions of Richmond (1974), Gabrielsen (1975), Schmidt (1976) and Greene (1980a). At the end of the 1970s and as an alternative to deterministic frontiers, stochastic frontiers appeared, allowing deviation in relation to the frontier to be also explained by a factor outside the firm's control (Lee and Tyler 1978). Stochastic frontier specification includes a two-sided error term, capturing not only the effects of the classical statistical noise but also technical inefficiency or the magnitude of the shortfall of output from its maximal possible value.

2.3.1 Deterministic Frontier Models

Under statistical deterministic models, all the deviations from the frontier are assumed to be the result of technical inefficiency of the production process, and no account is taken of measurement errors or statistical noise. The error term is completely due to inefficiency which may be specified according to a given asymmetric distribution of probability, such as a semi-normal, a truncated normal, an exponential (Schmidt 1976) or a gamma distribution (Greene 1990).

Aigner and Chu (1968) were the first authors to estimate a parametric and deterministic frontier model through a Cobb–Douglas function, which would express behaviour's heterogeneity of firms in the production context. According to them, differences captured from a cross-sectional group of units would be explained by technological reasons associated to the industry where they operate, by the scale of operations and by different options of management structures. When a firm operates at the frontier of production it is expected a zero disturbance in its model specification, meaning that it operates at the potential level of production.

By contrary, when a firm operates under the potential level of production, it is expected that the disturbance captures this fact, through a distance from the frontier, or the inefficiency magnitude. A deterministic parametric frontier may be specified as $Y_i = f(X_i; \beta)\mathrm{TE}_i$, where *i* indicates the producer, *Y* the scalar output, *X* a vector of inputs, $f(\cdot)$ the production frontier, β the parameter vector and TE technical efficiency. This last formulation is obtained through the ratio of the observed output to the maximum feasible output, or

$$\mathrm{TE}_i = \frac{Y_{i\,(\text{observed})}}{f\left(X_{i\,(\text{frontier})}, \beta\right)} \tag{2.1}$$

This formulation suggests that technical efficiency is assessed, for each productive unit, through $\mathrm{TE}_i = \exp(-ui)$, with $0 < \mathrm{TE_i} \leq 1$, reflecting the distance of each unit from the production frontier. So, the deterministic frontier formulation may be expressed through:

$$y_i = f\,(X_i;\beta)\exp(-u_i) \quad \mu_i \geq 0 \tag{2.2}$$

where y_i represents the dependent variable and translates the production observed for a productive unit *i*, with $i = 1, 2, \ldots\ldots N, \beta$ represents a vector of unknown technological parameters, x_i represents a vector of non-stochastic productive factors for observation *i* and u_i is the model's error component translating technical inefficiency and restricted to be $\geq$0, in order to guarantee that TE $\leq$ 1.

$$\ln y_i = \beta_0 + \sum_{n=1}^{N} \beta_n \ln x_{ni} - u_i$$

$$u_i \geq 0 \tag{2.3}$$

where the observations in u for each productive unit are non-negative, independent and identically distributed random variables, with an expected value to be positive and constant and the variance finite. It is also assumed that these random variables are not correlated with the regressors. Although the slope parameters in the deterministic frontier models can be consistently estimated by ordinary least squares (OLS) method, the constant term cannot be consistently estimated as the error term is not normality distributed. Considering some additional hypotheses or including adjustments on the specification of the error term of the frontier model, efficiency measures can also be obtained just from OLS as the adjustments restrict the frontier specification to be similar to the classical regression model.

Two methods for evaluating efficiency from OLS estimator involve the corrected ordinary least squares (COLS) and the modified least square (MOLS) as the estimated average-practicing frontier had been shifted up by the maximum amount of residuals (Gabrielsen 1975) or by the mean of residuals (Richmond 1974). Proposed by Winsten (1957) and developed by Gabrielsen (1975) and Greene (1980a, b) to estimate the frontier, COLS method adjusts the OLS line upwards or downward, by the maximum or by the minimum of the residuals, for a production function or for a cost function, respectively, in a way that COLS line rests parallel to the OLS line. Bounding all the units from above, inefficiency may be measured as a distance function from the COLS line so that all the unknown parameters may be consistently estimated.

In a first stage, OLS is used to obtain estimates of the slope parameters and a consistent but biased estimate of the intercept. In the second step, the estimated intercept is shifted up by the maximum value of the OLS residuals $\hat{\beta}_{\text{COLS}} = \hat{\beta}_{\text{OLS}} + \text{máx}\,\hat{e}_i$ so that the resulting COLS intercept is consistently obtained. Individual efficiency measures result from subtracting to an individual OLS residual the maximum sample residual such as $\hat{u}_i = \hat{e}_{\text{ols};i} - \max\,\hat{e}_{\text{ols};i}$.

Under a parametric context and proposed by Afriat (1972) and by Richmond (1974),[1] the modified least-squares method (MOLS) represents an alternative technique to OLS procedure in the frontier context, consisting in the correction of the intercept with the expected value of the error term such that the estimated frontier function could be displaced upwards by the estimate of $E[u_i]$, or, $\hat{u}_i$. As in the context of the COLS method, the residuals also provide consistent estimates of individual measures of efficiency, provided the estimated mean of the error term is subtracted: $-\hat{u}_i = \hat{e}_i - \hat{e}_i - \hat{u}_i$ instead of the maximum residual.

Indeed, this difference between two methods does not ensure that all units under MOLS estimation procedure are bounded from above by the estimated production frontier, although both frontiers (COLS and MOLS) led to parallel line to the OLS regression. The inconvenient of this parallelism is associated to the fact that it restricts the structure of the production technologies, best practice (frontier models) and mean practice (classical mean regression) to be equal. Also in the deterministic

[1]These two authors proposed an exponential and a half-normal to model the error term of the model, respectively.

frontier context, Greene (1980b) propose maximum-likelihood (ML) estimation, which represents the most popular and widely used technique in the estimation of frontier models. Kumbhakar and Lovell (2000) present an excellent review of the various distributions for modelling technical efficiency, necessary for maximum-likelihood estimation.

The main advantage of deterministic statistical models is the ease of obtaining individual estimates of efficiency for productive units although the estimation of a deterministic frontier, common to all productive units, assumes that all the deviations from the frontier are entirely interpreted as inefficiency. Indeed, in a deterministic frontier model context, maximum production is given by a function whose error term only reflects the firm's technical efficiency.

However, there are other factors outside its control which affect its behaviour and which are also 'captured' by the unilateral error term. So, the residuals of estimation provided by deterministic methods are therefore overvalued. Assessment of the performance of productive units and their comparison from efficiency measurement based on these residuals is consequently harmed. Summarizing, a deterministic and statistical frontier of production means that all observations (except one) are situated below the production frontier (or in the case of a cost function, above the cost frontier).

This restriction is the main limitation of using deterministic frontiers. The non-existence of a symmetric component in the error term able to capture random or uncontrollable shocks is the principal criticism of statistical deterministic frontier models.

2.3.2 *Stochastic Frontier Models*

The origins of parametric stochastic frontier analysis are in the efforts made to overcome the limitations imposed by deterministic models in assessing efficiency. This new approach assumes that the frontier varies randomly between productive units, that is, incorporating in its specification an additional error term that captures the effects caused by factors outside the productive unit's control. With a stochastic frontier to model an economic process, the error term is structured according to two components: a first component that would capture statistical noise and a second component that would translate the effects of technical inefficiency.

Both components of error term are specified from probability distributions (an asymmetric one for modelling inefficiency component and a symmetric/normal distribution for modelling the stochastic error component). From an econometric perspective, the stochastic statistical method refers to estimation of models based on functional forms that allow observations on both sides of the frontier.

This method was proposed almost simultaneously in three continents: Meeusen and Van den Broeck (June 1977), Aigner et al. (July 1977) and Battese and Corra (1977), in an attempt to overcome the disadvantages caused by deterministic frontiers in assessing the individual efficiency of productive units.

These models possess not only a component reflecting the distance from the frontier due to producers' technical inefficiency but also a new component that absorbs the impact of random shocks on production. These models are characterized by having an error term with two elements $\varepsilon = v - u$. The first component of the error, usually assumed to follow a Gaussian distribution, reflects disturbances or factors which affect the production level but cannot be controlled (climatic variations, equipment breakdown, illnesses . . .).

The second component of the error captures the inefficiency in managing production and is assumed to follow a one-sided distribution, such as an half-normal (Aigner et al. 1977), a truncated normal (Stevenson 1980), an exponential (Meeusen and van den Broeck 1977) and a gamma (Greene 1990). Considering a stochastic formulation for a frontier model, the following Cobb–Douglas log-linear function will include a compound error term and a deterministic part as is showed in (2.4):

$$\ln Y_i = \beta_0 + \sum_{n=1}^{N} \beta_n X_{ni} + v_i - u_i \tag{2.4}$$

where Y_i is the logarithm of production concerning producer $i, i = 1, \ldots N$, X_i is a vector of productive factors used by producer i, β is a vector of technological parameters to be estimated, (v_i) are random i.i.d variables with zero mean and independent of u_i and of the regressors and $u_i \geq 0$ are non-negative random i.i.d variables, independent of v_i and of the regressors. Both error components, v and u, have constant means (0 and μ) and variances, σ_v^2 and σ_u^2, respectively, over all observations.

The joint density of these two error components will underlie likelihood functions. So, according to the distribution assumed for the asymmetric component of the error term, model designations will be given by normal-half-normal, normal-truncated-normal, normal-exponential and, finally, normal-gamma. Initially, stochastic frontiers only allowed estimation of one measurement for the sample's average efficiency.

> The main limitation of the stochastic frontier model is the impossibility of separating the two components from the individual residual, i.e., it does not allow estimation of individual technical inefficiency. (Forsund et al. 1980)

Two years later, Jondrow et al. (1982) presented a method which was able to overcome this major limitation of stochastic frontiers. They showed that for the half-normal case, the expected value of u_i conditional on the composed error term is

$$E\left[\frac{u_i}{\varepsilon_i}\right] = \frac{\sigma\lambda}{(1+\lambda^2)}\left[\frac{\phi(e_i\lambda/\sigma)}{\Phi(-e_i\lambda/\sigma)} - \frac{e_i\lambda}{\sigma}\right] \tag{2.5}$$

where $\phi(\cdot)$ represents the density of the standard normal distribution and $\Phi(\cdot)$ the cumulative density distribution,

$$\lambda = \frac{\sigma_u}{\sigma_v}, \quad e_i = v_i - u_i \quad \text{and} \quad \sigma = \left(\sigma_u^2 + \sigma_v^2\right)^{1/2} \tag{2.6}$$

The basic idea of this method consisted in using the mean and mode of conditional distribution to obtain estimates for each producer as $\mathrm{TE}_i = 1 - E\left[u_i/e_i\right]$. Jondrow et al. (1982) applied the methodology to models with the u_i component specified from a semi-normal and an exponential distribution. For the models described through a Cobb–Douglas function and in the form of logarithm, Battese and Coelli (1988) suggested the use of another estimator adapted to sectional data and for normal-truncated-normal and for normal-half-normal distributions and expressed as

$$E\left[\exp\frac{(-u_i)}{e_i}\right] = \left[\frac{1-\Phi\left(\delta + (\gamma e_i/\delta)\right)}{1-\Phi(\gamma e_i/\delta)}\exp\left(\gamma e_i + \left(\frac{\delta^2}{2}\right)\right)\right] \tag{2.7}$$

where

$$\delta = \frac{\sigma_u \sigma_v}{\sigma}; \quad \gamma = \frac{\sigma_u^2}{\sigma^2}. \tag{2.8}$$

With sectional data, two methods of estimating stochastic frontiers are generally analysed: maximum likelihood (ML) – Afriat (1972), Greene (1980b) and Stevenson (1980); and modified least squares (MOLS). The option between methods based on OLS and the ML method also depends on the distribution asymmetry intensity of u_i: when this is not very pronounced, distribution of the error term is approximately symmetric and normal. In these circumstances, results of estimation are similar to those obtained with OLS. The efficiency gains attained by using the ML method only occur if the joint distribution of the error term is clearly asymmetric.

Calculation of the sample's mean technical efficiency can be made from the mean of the residuals from stochastic model estimation: with v_i being a random variable normally distributed with zero mean, the value of the sample's mean efficiency is identical to the mean of the asymmetric component of the error or $E\left[\varepsilon_i\right] = E\left[v_i - u_i\right] = E[-u_i]$. So, $\overline{\mathrm{ET}} = 1/N\left[\sum_i(-\hat{u}_i)\right]$ or $\overline{\mathrm{ET}} = 1/N\left[\sum_i \exp(-\hat{u}_i)\right]$ if the model is presented in logarithmic form.

The expressions for the sample's mean efficiency and for the respective expected value depend on the probability distribution assumed for u_i (Jondrow et al. 1982). Therefore, if a semi-normal or exponential distribution is assumed, the sample's mean efficiency will be given by $\overline{\mathrm{ET}} = -\hat{\sigma}_u\sqrt{2/\pi}$ and by $\overline{\mathrm{ET}} = 1/\hat{\gamma}$ respectively.

The main disadvantages arising from estimating stochastic frontiers from sectional data for assessment of technical efficiency of productive units are associated with the requirement of major restrictions, such as the absence of correlation between regressors and the term of technical efficiency, suitability of the chosen distributions for modelling the asymmetric component of the compound error and the impossibility of ensuring consistency of the estimators when a productive unit is observed only once. The variance of distribution assumed for the component of technical efficiency conditional in the entire error term does not disappear when the size of the sample increases. Advantages of the panel data stochastic frontier models versus cross-sectional data are presented and explored in Schmidt and Sickles (1984).

2.4 Frontier Models in Panel Data Framework

The literature is very rich with regard to theoretical and empirical use of data set in a panel (see Baltagi 1995). The first parametric approach to frontier models with panel data for estimating measurements of technical inefficiency is due to Pitt and Lee (1981), who combined the potential of analysing time series with the advantages of sectional analysis to estimate frontier models through maximum likelihood. This was followed immediately by innovative studies in this domain, responsible for the later development of this type of model.

The contributions of Schmidt and Sickles (1984), Cornwell et al. (1990), Battese and Coelli (1988, 1992, 1995) and Kumbhakar et al. (1991) stand out particularly. For a review of the literature on using panel data in the context of stochastic frontiers, see the studies by Kumbhakar (1990) and Kumbhakar and Lovell (2000). In the literature referring to frontier models estimated from panel data, we find different econometric specifications, resulting from the various hypotheses assumed for the term of technical inefficiency (Ahmad and Bravo-Ureta 1996). Regarding the hypotheses assumed for the term of technical inefficiency, these hypotheses can be summarized as follows:

(a) Absence of correlation between the term of efficiency and regressors
(b) Correlation between the term of efficiency and regressors
(c) Temporal invariance of the term of technical efficiency
(d) Temporal variance of the term of technical efficiency

The models are classified according to the hypotheses assumed for the term of technical inefficiency and according to the method of estimation adopted. The following categories may be therefore considered:

A – Models with time-invariant inefficiency term

- Fixed effect models
- Random effect models
- Maximum-likelihood models

B – Models with time-variant inefficiency term

- Models based on least-squares correction (fixed effect model and random effect model)
- Maximum-likelihood models
- Exogenous influence function models

2.4.1 Fixed Effect Model with Time-Invariant Inefficiency Term

In the non-frontier context of estimation with panel data, the fixed effect model was introduced by Mundlack (1961) and developed by Hock (1962) among others.

As the starting point, it is consider the general model (Cobb–Douglas), of just one product, for panel data:

$$\ln y_{it} = \beta_0 + \sum_{n=1}^{N} \beta_{nit} \ln x_{nit} + \varepsilon_{it} \tag{2.9}$$

with $i = 1, \ldots, N$ producers; $t = 1, \ldots, T$ periods and $n = 1, \ldots N'$, explanatory variables. The level of production for individual i in period t is represented by $\ln y_{it}$, the independent term is given by β_0 and the β_{nit} regression coefficients may or may not vary in i or in t. It is noted that in a traditional panel data model, the error term has an expected value of $E[\varepsilon_{it}] = 0$ and a constant variance $V[\varepsilon_{it}] = \sigma_\varepsilon^2$. Schmidt and Sickles (1984) proved that, in a context of using panel data to estimate efficiency measurements, it was possible not to specify a particular distribution for the effects of inefficiency, since the model's parameters could be estimated using traditional estimation methods with panel data, where the fixed effect method is included.

Schmidt and Sickles (1984) considered models where individual effects are constant parameters that can be correlated with the explanatory variables. Coefficients are estimated from the idea of variation in productive units, within-firm variation (Farsi et al. 2005, 2006), not being affected by the existence of correlation between regressors and individual effects. The specification underlying this type of model suggests that the differences found in terms of productive factors are simply explained by a set of individual factors, constant over time, which in frontier models translate technical inefficiency. Productive structure is identical for all firms. The estimation techniques adopted depend on the absence, or not, of correlation between regressors and the technical inefficiency term and on the imposition, or not, of a specific distribution for the technical inefficiency term.

The frontier model with time-invariant technical efficiency is therefore expressed as follows:

$$y_{it} = \beta_0 + \sum_{n} \beta_n \ln x_{nit} + v_{it} - u_i \quad \mu_i \geq 0 \tag{2.10}$$

where $i = 1, \ldots, N$, $t = 1, \ldots, T$, $n = 1, \ldots, N'$ are, respectively, the index distinguishing the different productive units, the time index and the index describing the N' regressors used by producer i. $v_{it} - u_i$ represents the compound term of model disturbance, where the first component of the error is an i.i.d random variable, $E[v_i] = 0$ and $V[v_i] = \sigma_v^2$, which varies over units and time and the second component (u_i) is the asymmetric inefficiency error term, assumed to vary only over units.

This last error component is treated as firm-specific constants or as fixed effects or individual effects. It is assumed that $u_i \geq 0$, that is, the non-negativity of this component for any i, and also that $E[u_i] = \mu$ and $V[u_i] = \sigma_u^2$. It is also considered that u_i is distributed independently of v_{it} and that v_{it} is not correlated with the regressors. The model presented in (2.10) may be adapted for OLS estimation proposes, by eliminating the intercept term and adding a dummy variable for each sample element:

$$y_{it} = \beta_{0i} + \sum_{n} \beta_n \ln x_{nit} + v_{it} \tag{2.11}$$

where $\beta_{0i} = (\beta_0 - u_i)$ represents the specific N intercepts associated with each producer. In this new model, named the fixed effect frontier model, no distribution is specified for the asymmetric error term (u_i) which can be correlated with the regressors or with v_i. As the component u_i is estimated together with the specific intercept of each producer, it is treated as a fixed effect (not random). OLS estimation procedure is applied, and individual estimates are obtained for $\hat{\beta}_{0i}$ from the mean of the within estimation residuals and by productive unit: $\hat{\beta}_{0i} = \overline{\hat{\varepsilon}_{\iota W}}$ where $\overline{\hat{\varepsilon}_{\iota W}} = \bar{y}_{\iota} - \beta' (x_{it} - \bar{x}_{\iota})$ with $\bar{y}_{\iota} = \sum_{t=1}^{T} y_{it}/T$ and $\bar{x}_{\iota} = \sum_{t=1}^{T} x_{it}/T$ and $\bar{v}_{\iota} = \sum_{t=1}^{T} v_{it}/T$. The estimate for u_i is obtained from the following correction (Gabrielsen 1975 and Greene 1980a method): $\hat{u}_i = \hat{\beta}_0 - \hat{\beta}_{0i}$. This correction ensures the positivity of the individual effects and is done on the assumption that the most efficient firm is 100 % efficient, that is, it presents when $\hat{\beta}_0 = \max \hat{\beta}_{0i}$. The estimates are consistent when the respective variances tend towards 0 as the number of observations ($N \times T$) tends towards infinity. Individualized estimates for the measurement of technical efficiency are given by $\mathrm{TE}_i = \exp\{-\hat{u}_i\}$ with $\hat{u}_i = \hat{\beta}_0 - \hat{\beta}_{0i}$, that is, it is given by the difference between the global estimate for the intercept and the estimates obtained for producers' specific intercepts.

The global estimate for the intercept is the result of $\hat{\beta}_0 = \max\left\{\hat{\beta}_{0i}\right\}$. The producer situated on the frontier is considered 100 % efficient presenting $\max\left\{\hat{\beta}_{0i}\right\}$, and the efficiency of the others is assessed in relation to this producer. Estimation with a fixed effect frontier model (within estimator) generates consistent estimates for the technological parameters $\hat{\beta}_n$ when T or N tends towards infinity, without the need to assume absence of correlation of the asymmetric error term with the regressors or normality of the symmetric error term distribution. As for the consistency property for the estimates of the specific intercepts for each producer, $\left\{\hat{\beta}_{0i}\right\}$, this is only found when $T \to \infty$. Estimates for the asymmetric error term are only consistent if N and $T \to \infty$. Another possible transformation to the previously specified model (2.10) consists in estimating a frontier model (2.12) by OLS after the within-groups transformation (or after all observations have been transformed in order to be expressed in terms of deviations from the individual means):

$$y_{it} - \overline{y_{it}} = \beta\prime (x_{it} - \overline{x_i}) + v_{it} - \overline{v_i} \tag{2.12}$$

The possibility of obtaining consistent estimators of individual technical efficiency, even faced with the hypothesis of correlation between regressors and individual effects, is the main attraction of estimation based on fixed effect models. Another relevant advantage lies in the fact of u_i is fixed, and so, the specification of the respective distribution is not necessary.

The main limitations arising from using these models can be expressed in the following points: (a) estimation of the model means the variables must present sufficient time variations, since the within technique assumes the respective parameters estimated from the deviations of variables from the respective means. If the variation is small, the accuracy of the estimates is in doubt; (b) the measurements of technical inefficiency reflect not only the inefficiency of the productive process but also the effect of other factors which are invariable over time and variable between productive units; (c) the productive unit with the smallest intercept is understood to be the efficient unit with which the other units are compared.

Among many other studies with fixed effect production frontier models, that of Ahmad and Bravo-Ureta (1996) stands out, comparing the effects on measuring the efficiency of fixed effect models and stochastic frontier models with different specifications for the asymmetric error component (semi-normal and truncated normal). It should be mentioned that the first fixed effect frontier models only considered balanced data panels. Some years later, these models were developed and adapted by Battese and Coelli (1988) to integrate also unbalanced panels.

2.4.2 *Random Effect Model with Time-Invariant Inefficiency Term*

The first developments in the sphere of random effect models were the work of Pitt and Lee (1981). The authors considered a model with distributional assumptions about the error term where $v_{it} \cap \text{i.d.}N\left(0,\sigma_v^2\right)$ represents noise and $u_{it} \cap \text{i.d.}N^+\left(0,\sigma_u^2\right)$ represents distribution of the non-negative component which translates the inefficiency of the model.

For the respective estimation, Pitt and Lee (1981) proposed the ML technique. Several years later, Battese and Coelli (1988) adopted this formulation, proposing truncated-normal distribution for modelling the component of technical inefficiency and using ML for estimation purposes. Schmidt and Sickles (1984) used another random effect model aiming to avoid the drawbacks arising from the Pitt and Lee specification which assumed a particular distribution for the inefficiency component and regressors variable over time. Assuming now independence of the inefficiency term and the regressors and that u_i are random than fixed results new modification of the initial model (2.10) or the random effect model given by the expression:

$$\ln Y_i = \beta_0^* + \sum_{n=1}^{N} \beta_n \ln X_{nit} + v_{it} - u_i^* \tag{2.13}$$

where $\beta_0^* = \beta_0 - E(u_i)$ and $u_i^* = u_i - E(u_i)$ and zero mean for u_i^* and v_i. With the introduction of this transformation, zero mean for the error term, GLS (generalized least squares) technique can be applied to estimate the model (2.13). The random effect model operates in exactly the same way as the error component (one-way) model described in the literature on panel data. To estimate this model, the GLS technique in two steps is used.

The method involves, at the first stage, OLS estimation of all the model's parameters. When the matrix of covariances of the error $v_{it} - u_i^*$ is known, that is, σ_v^2 and σ_u^2 are known, the GLS estimator for β_0^* and for β_n is BLUE (best linear unbiased estimator), and consistency is ensured either when $N \to \infty$ or when $T \to \infty$. However, usually σ_v^2 and σ_u^2 are not known. In this situation, it is appropriate to use the FGLS (feasible generalized least squares) method to estimate the variance of the compound error term $\hat{V}[\varepsilon] = \hat{V}[u_i] + \hat{V}[v_i] = \hat{\sigma}_u^2 + \hat{\sigma}_v^2$. The estimate for the variance of the symmetric error term is given by the variance of the residuals of the fixed effect model (within[2] residuals), $\hat{\sigma}_v^2 = \hat{\varepsilon}'\hat{\varepsilon}/[N(T-1)-K]$ and the estimate for the variance of the asymmetric error term is given by the combination of the residuals of the between[3] estimation with the residuals of the within estimation $\hat{\sigma}_v^2 = \{\hat{\varepsilon}'\hat{\varepsilon}/[N-K] - \hat{\sigma}_v^2\}/T$. At a second stage and after estimation of β_0 and β_n (with GLS or FGLS), the measurements of technical efficiency are given by $\text{TE}_i = \exp\{-\hat{u}_i\}$ with $\hat{u}_i = \max\{\hat{u}_i^*\} - \hat{u}_i^*$ and u_i^* resulting from the mean residuals of FGLS estimation:

$$\hat{u}_i^* = \frac{1}{T}\sum_t \left(\ln y_{it} - \hat{\beta}_i^* \sum_n \hat{\beta}_n \ln x_{nit}\right)^4 \tag{2.14}$$

In these conditions, the estimates obtained for individual inefficiency translate, just as in the case of the fixed effect model, the distances between the intercept of each productive unit and the greatest intercept relating to the productive unit considered efficient. The frontier is then moved to the greatest intercept estimated in the sample. The BLUP (best linear unbiased predictor) by Lee and Griffiths (1979) is an alternative estimator to u_i^* and is given by

$$\tilde{u}_i^* = \frac{-\hat{\sigma}_u^2}{T\hat{\sigma}_u^2 + \hat{\sigma}_u^2}\sum_{t=1}^{T}\left(\ln y_{it} - \hat{\beta}_{0\,\text{FGLS}}^* - \hat{\beta}_{n\,\text{FGLS}} \ln x_{it}\right) \tag{2.15}$$

[2]The residuals of the within estimation are given by $\hat{\varepsilon}'\hat{\varepsilon} = \sum_{i=1}^{N}\sum_{t=1}^{T}\left[y_{it} - \overline{y_i} - \hat{\beta}'_{\text{Within}}(x_{it} - \overline{x_i})\right]^2$.

[3]$\varepsilon^{*\prime}\varepsilon^* = \sum_{i=1}^{N}\left(\overline{y_i} - \hat{\beta}^*_{\text{Between}} - \hat{\beta}'_{\text{Between}}\overline{x_i}\right)^2$. The latter residuals are the result of applying the OLS technique to the model: $\overline{y}_{it} = \beta^* + \beta'\overline{x_i} + \overline{v_i} - \overline{u_i^*}$.

[4]These estimates are consistent as long as N and $T \to \infty$.

The GLS[5] estimator for β_0^* and for β_n^* is consistent when simultaneously N and $T \rightarrow \infty$, and the variances of the two components of the error term are known. When these are unknown, it is necessary that $T \rightarrow \infty$ for the variance of u to be estimated consistently and that N or $T \rightarrow \infty$ for the FGLS estimator of the variance of v to be consistent. The estimators of $\tilde{u}_i$ and $\hat{u}_i$ are consistent when N and $T \rightarrow \infty$. The FGLS estimator is suitable when N is large and when the hypothesis of the existence of correlation of u with the regressors is rejected. Opting for the FGLS estimator or the within estimator depends on the hypothesis of absence of correlation between technical inefficiency and the regressors being confirmed or not.

Hausman and Taylor (1981) developed an alternative estimator (HT) which shares characteristics with within and FGLS estimators, being indicated for testing this hypothesis of absence of correlation between technical inefficiency and the regressors. Equally, adoption of a random effect model also involves limitations. Firstly, these models do not allow us to distinguish the inefficiency of non-observable heterogeneity (only one parameter is specified to capture that heterogeneity).

As in the fixed effect model context, to check consistency of the asymmetric error term, it is necessary that $T_i \rightarrow \infty \forall_i$. Secondly, the frontier associated with a random effect model is constructed from moving the frontier to the intercept relating to the most efficient unit in the sample. However, if the sample is small, it may not include any efficient unit or one close to optimal production.

2.4.3 Temporal Variation of Efficiency Term

The fixed effect, random effect and maximum-likelihood models share the assumption of temporal invariance in the component of technical inefficiency. However, we find that, when analysing the efficiency of a productive process, it is often more appropriate to consider the time effect on this component of the error, principally when there are sufficient data on the same productive unit observed in various periods. In these circumstances, it is improbable that productive units continue to present a constant measurement of inefficiency in all the periods of observing their production.

Indeed, knowledge of the level of technical inefficiency over time necessarily causes interventions in the production process that affect results in the following periods, invalidating the initial hypothesis of individual effects being constant in t. Cornwell et al. (1990), Lee and Schmidt (1993), Heshmati and Kumbhakar (1994),

[5]When the variances of the components of the error term are unknown, the FGLS estimator of σ_v^2 is consistent with N or $T \rightarrow \infty$, whereas for σ_u^2 consistency is ensured only with $T \rightarrow \infty$.

Kumbhakar and Heshmati (1995) and Battese and Coelli (1992, 1995) were pioneers in studying frontier models with time variation of technical efficiency, and Cuesta (2000) enhanced the literature on this subject with new developments.

The different models, which in the respective specifications integrate the time variation of technical efficiency, can be classed in three categories: (1) fixed effect models and random effect models (based on correction of least squares), (2) models estimated using ML and (3) exogenous influence function models. The main difference between the categories of models with time-variable efficiency lies in how this is modelled. In fixed and random effect models, efficiency is modelled from the independent term, whereas in maximum-likelihood models and those incorporating exogenous influences, this is modelled from the error term, therefore meaning imposition of joint probability distribution.

For each of the first two categories mentioned, they are also differentiated by the specifications where the standard of time-variable efficiency is common to all producers from those specifications where the efficiency standard varies between producers. Prominent among the first type of model are those by Cornwell et al. (1990), with a distinct standard of time variation for each productive unit,[6] and by Lee and Schmidt (1993), with a standard of time variation common to all productive units, both estimated using COLS techniques. In these models, it is not possible to separate the effect of technological progress from the effect of technical inefficiency in the independent term.

In the context of the second group of models, assuming simultaneously the conditions of independence of u regarding the regressors and the distributional suppositions mentioned concerning the two components of the respective errors, this includes those satisfying the conditions required for maximum-likelihood estimation. Of special note are the models suggested and developed by Kumbhakar (1990), Battese and Coelli (1992, 1995) and Cuesta (2000). While the models by Kumbhakar (1990) and Battese and Coelli (1992) have a standard of time-variable efficiency common to all firms, the models proposed by Battese and Coelli (1995) and Cuesta (2000) present a rate of time-variable inefficiency which varies between firms. In this second group of models, it is possible to separate the effect of technological progress from the effect of inefficiency on the productive process.

The third category of models concerns a type of frontier model where the term of inefficiency is a vector of observable factors. Kumbhakar et al. (1991) developed a model in a sectional context, and Battese and Coelli (1995) generalized it for panel data context.

[6]The effects where the inefficiency is contained are given by the product of time effects (common to all firms) and individual effects: $\beta_{it} = \theta_t \delta_i$.

2.4.3.1 Fixed and Random Effect Models

According to the literature on the subject, in this type of model, time-variable efficiency is incorporated from the independent term also named individual effects. In these models, the independent term incorporates two components: one reflecting the effect of technological progress and the other the effect of technical inefficiency on the productive process. Concerning the estimation method, as in the case of time-invariant inefficiency, this is based on corrections carried out on the results of estimation obtained with the least-squares method.

Two subgroups of models are distinguished, however: those for which a standard of efficiency variation is assumed for each productive unit and models where this standard is common to all productive units. The Cornwell et al. (1990) model is included in the first type of models, and the model proposed by Lee and Schmidt (1993) falls into the second type.

The Cornwell et al. (1990) Model

Cornwell et al. (1990) developed and estimated a frontier model in whose specification they introduced the component of technical inefficiency varying over time. The functional form is based on a Cobb–Douglas technology of production where the independent term and some regression coefficients vary with the individuals and with time.

This model generalizes the one by Schmidt and Sickles (1984), considering in the production function, a function of variable coefficients which is quadratic in t and contemplates the individual and time variation of technical efficiency which is identical for all productive units. The model can therefore be presented with the following formulation:

$$\begin{aligned} y_{it} &= \beta_{0t} + \sum_k \beta_n \ln x_{nit} + v_{it} - u_{it} \\ &= \beta_{it} + \sum_k \beta_n \ln x_{nit} + v_{it} \end{aligned} \tag{2.16}$$

with $\beta_{it} = \beta_{0t} - u_{it}$ expressing the intercept of unit i in period t and with β_{0t} translating the common frontier intercept in period t. The authors assume a quadratic function to explain, or $\mu_{it} = \Phi_{1i} + \Phi_{1i}t + \Phi_{3i}t^2$, where Φ's are specific producer parameters. With this new specification, production levels vary between firms and over time, as well as technical efficiency. For estimation proposes, authors applied GLS random effect estimator, as they assumed time-invariant regressors, time-varying technical efficiency in the specified model and independence between the asymmetric error term and regressors.

Lee and Schmidt (1993) Model

The main difference in relation to the last model lies in the fact that these authors have imposed a standard time variation for u_i, which is identical for all the sample's productive units. They propose an alternative specification for the asymmetric error component $\mu_{it} = \delta_t \mu_i$ or the product between the time effects (dummies) and individual producers' inefficiencies.

2.4.3.2 Maximum-Likelihood Models

For a second group of models, there is simultaneous assumption of the conditions of independence of u_i with regard to the regressors and distributional suppositions about the two components of the respective errors. These models can then be estimated by ML, highlighting the models suggested and developed by Kumbhakar (1990) and Battese and Coelli (1992), with a standard of time variation common to all producers, and by Battese and Coelli (1995) and Cuesta (2000), where the standard time variation is specific for each producer.

Kumbhakar (1990) Model

Kumbhakar (1990) was the first to suggest and use a stochastic frontier model with a time variation standard of levels of technical inefficiency common to all productive units with estimation by maximum likelihood. The author suggests that the time variation of u assumption may be defined through an exponential function of time given by $u_{it} = f(t) \cdot u_i$ or the product of a function in t or

$$f(t) = \left[1 + \exp\left(\alpha t + \beta t^2\right)\right]^{-1} \tag{2.17}$$

by u_i which is modelled as a truncated-normal distribution $u_i \cap$ i.i.d.$N^+(0, \sigma_u^2)$, independent of regressors. The estimation of the time-varying efficiency effect model is realized in a random effect framework and using ML method.

Battese and Coelli (1992) Model

Battese and Coelli (1992) generalized Kumbhakar's idea to unbalanced (or incomplete panel data) models, proposing an alternative frontier model to that of Kumbhakar, also assuming time varying for efficiency and restricted to be common to all individuals (productive units).

The author suggests that the time variation of u should be defined from a function given by $u_{it} = f(t) \cdot u_i$, where

$$f(t) = \exp\left[\eta\,(t - T)\right], \quad f(t) \geq 0; \quad f(T) = 1 \tag{2.18}$$

involving only one unknown parameter and so, less flexible (Coelli et al. 1998, pp. 278). Since the rate of time variation is identical for all units, the value estimated for the additional parameter allows analysis of the tendency of efficiency over time: efficiency increases (for all units) if the value estimated for this parameter is positive, diminishing if it is constant over time if the parameter assume a zero value.

With this new specification only in the last year of analysis, productive units present a specific standard of efficiency. In the other years, the standard is also common to all productive units. Battese and Coelli (1992) used the method of maximum likelihood together with a truncated-normal distribution for the asymmetric error term modelling.

2.4.3.3 Models with Exogenous Influences

The third category of models is used as the result of research for an answer to the existence of inefficiency, and it is associated with a type of frontier where the inefficiency term is a vector of observable factors. Kumbhakar et al. (1991) developed this methodology in the context of sectional data, and Battese and Coelli (1995) generalized it to panel data.

For Kumbhakar et al. (1991), technical inefficiency would be composed by a deterministic part with exogenous variables and by a stochastic component $u_i = \gamma' m_i + \theta_i$ with m_i being a vector of observable qualitative factors, γ a vector of parameters and non-observable random component (inefficiency model error term).

Additionally, Kumbhakar et al. (1991) assuming (1) $v_i \cap$ i.i.d. $N(0, \sigma_v^2)$, (2) $u_i \cap N^+\left(\gamma' m_i, \sigma_u^2\right)$ and (3) v and u independently distributed, suggested ML estimation procedure. According to this specification, technical inefficiency only varies between productive units and depends on specific exogenous variables.

This type of model has been criticized due to the fact that it tries to explain differences in inefficiency through variables that already appeared in the models, as productive factors.

Battese and Coelli (1995) Model

Battese and Coelli proposed a model where the inefficiency term varies over time, follows truncated-normal distribution and is a function of certain explanatory variables, such as $u_{it} = z_{it}\delta + W_{it}$. δ is a unknown vector of coefficients, z represents a vector of observable explanatory variables and W the error term of the inefficiency model.

These explanatory variables (also called exogenous variables or inefficiency effects) of the inefficiency effects model include determinant factors for understanding the magnitude of the distance of the observed production in relation to the corresponding production situated on the respective stochastic frontier. The term

(ε) or error term of the Battese and Coelli (1995) model is given by two random variables, $\varepsilon_{it} = v_{it} - u_{it}$, which translate two types of effects or disturbances affecting the productive process.

The first component is the error term that captures effects caused by errors of measurement and by all factors outside the productive unit's control and is modelled through a Gaussian (0, 1). The second component is a non-observable measurement of technical inefficiency which may be time varying and controllable by the unit. This component measures the magnitude of the effort made to reduce the distance from the technological frontier.

These variables are therefore non-negative variables representing the technical inefficiency of production. Values for the inefficiency measurements are derived from generalization of the Jondrow et al. (1982) methodology, that is, deduced by means of the expected conditional value of the estimated error value, and the other parameters of the model are estimated in just one step using maximum likelihood.

The size of the panel has quality estimation implications whatever the method used. The most favourable situation, or when N and $T \to \infty$, allows consistent estimators. When T is high, but N is low, the within estimator subject to hypotheses of correlation between regressors and technical inefficiency and absence of distribution for u_i is the most appropriate. When N is high, but T is low, consistency is not guaranteed whatever the method used.

2.5 New Developments

When firm characteristics are not taken in account and they are erroneously estimated as being inefficiency, it may cause serious biases in efficiency results. Modelling heterogeneity among organizations represents an important field of research. Indeed, in the conventional panel data context, firm-specific heterogeneity was incorrectly considered as inefficiency.

In the fixed effects model of Schmidt and Sickles (1984), time-invariant unobserved heterogeneity was captured by the inefficiency component until Greene's (2005) suggestion of the true fixed effects model which restricts fixed effects to only represent the unobserved firm heterogeneity and not inefficiency. Research in this field has been conducted by Farsi et al. (2005, 2006) and Greene (2005). Modelling heterogeneity has been extended to the Bayesian context by Caudill et al. (1995), Tsionas (2001, 2002) and Huang (2004).

Nowadays, stochastic frontier approach to organizations efficiency measurement may play an important role in the field of economical sustainable development of societies, as it allows managements to reach high levels of performance restricting resources, minimizing undesired outputs or optimizing desired outputs. More research is needed in order to enlarge the knowledge in this issue.

References

Afriat SN (1972) Efficiency estimation of a production function. Int Econ Rev 13(3):568–598

Ahmad M, Bravo-Ureta BE (1996) Technical efficiency measures for dairy farms using panel data: a comparison of alternative model specifications. J Product Anal 7:399–415

Aigner D, Chu SF (1968) On estimating the industry production function. Am Econ Rev 58:826–839

Aigner D, Lovell CAK, Schmidt P (1977) Formulation and estimation of stochastic frontier production function models. J Econ 6:21–37

Arrow K, Chenery H, Minhas B, Solow R (1961) Capital labor substitution and economic efficiency. Rev Econ Stat 45:225–247

Baltagi B (1995) Econometric analysis of panel data. Wiley, New York

Battese GE, Coelli TJ (1988) Prediction of firm-level technical efficiencies with a generalised frontier production function and panel data. J Econ 38:387–399

Battese G, Coelli T (1992) Frontier production functions, technical efficiency and panel data: with application to paddy farmers in India. J Product Anal 3:153–169

Battese G, Coelli T (1995) A model for technical inefficiency effects in a stochastic frontier production function for panel data. Empir Econ 20:325–332

Battese G, Corra G (1977) Estimation of a production frontier model: with application to the pastoral zone of eastern Australia. Aust J Agric Econ 21(3):169–179

Bhattacharyya A, Parker E, Raffiee K (1994) An examination of the effect of ownership on the relative efficiency of public and private water utilities. Land Econ 70(2):197–209

Caudill S, Ford J, Gropper D (1995) Frontier estimation and firm specific inefficiency measures in the presence of heteroscedasticity. J Bus Econ Stat 13:105–111

Charnes A, Harnes A, Cooper WW, Rhodes E (1978) Measuring the efficiency on decision making units. Eur J Oper Res 2:429–444

Cobb S, Douglas P (1928) A theory of production. Am Econ Rev 18:139–165

Coelli T, Prasada Rao DS, Battese GE (1998) An introduction to efficiency and productivity analysis. Kluwer Academic Publishers, Boston

Cornwell C, Schmidt P, Sickles RC (1990) Production frontiers with cross-sectional and time-series variations in efficiency levels. J Econ 46(1/2):185–200

Cuesta RA (2000) A production model with firm-specific temporal variation in technical inefficiency: with application to Spanish dairy farms. J Product Anal 13:139–158

Dean J (1951) Managerial economics. Prentice Hall, Englewood Cliffs

Debreu G (1951) The coefficient of resource utilization. Econometrica 19:273–292

Farrell M (1957) The measurement of productive efficiency. J R Stat Soc A Gen 120:253–281

Farsi M, Filippini M, Greene WH (2005) Efficiency measurement in network industries: application to the Swiss railway companies. J Regul Econ 28(1):69–90

Farsi M, Filippini M, Kuenzle M (2006) Cost efficiency in regional bus companies: an application of alternative stochastic frontier models. J Transp Econ Policy 40(1):95–118

Forsund FR (1992) A comparison of parametric and non-parametric efficiency measures: the case of Norwegian ferries. J Product Anal 3:25–43

Forsund FR, Hjalmarsson L (1979) Generalised Farrell measures of efficiency: an application to milk processing in Swedish dairy plants. Econ J 89:294–315

Forsund FR, Sarafoglou N (2002) On the origins of data envelopment analysis. J Product Anal 17(1–2):23–40

Forsund F, Lovell K, Schmidt P (1980) A survey of frontier production functions and of their relationship to efficiency measurement. J Econ 13:5–25

Gabrielsen A (1975) On estimating efficient production functions. Working Paper No. A-85, Chr. Michelsen Institute, Department of Humanities and Social Sciences, Bergen, Norway

Greene W (1980a) On the estimation of a flexible frontier production model. J Econ 3:101–115

Greene W (1980b) Maximum likelihood estimation of econometric frontier functions. J Econ 13:27–56

Greene W (1990) A gamma distributed stochastic frontier model. J Econ 46:141–163
Greene W (1997) Frontier production functions. In: Pesaran MH, Schmidt P (eds) Handbook of applied econometrics, vol II, Microeconomics. Blackwell, Oxford
Greene W (2005) The econometric approach to efficiency measurement. In: Lovell K, Fried H, Schmidt S (eds) The measurement of productive efficiency, 2nd edn. Oxford University Press, New York
Hausman JA, Taylor WE (1981) Panel data and unobservable individual effects. Econometrica 49:1377–1398
Heshmati A, Kumbhakar SC (1994) Farm heterogeneity and technical efficiency: some results from Swedish dairy farms. J Product Anal 5:45–61
Hock I (1962) Estimating of production function parameters combining time series and cross-section data. Econometrica 30:1
Huang HC (2004) Estimation of technical inefficiencies with heterogeneous technologies. J Product Anal 21:277–296
Johnston J (1959) Statistical cost analysis. McGraw-Hill, New York
Jondrow J, Lovell K, Materov I, Schmidt P (1982) On the estimation of technical inefficiency in the stochastic frontier production function model. J Econ 19:233–238
Kalirajan KP, Shand RT (1999) Frontier production functions and technical efficiency measures. J Econ Surv 13(2):149–172
Koopmans TC (1951) An analysis of production as an efficient combination of activities. In: Koopmans TC (ed) Activity analysis of production and allocation. John Wiley and Sons, Inc.
Kopp RJ, Diewert W (1982) The decomposition of frontier cost function deviations into measures of technical and allocative efficiency. J Econ 19(2/3):319–332
Kumbhakar S (1990) Production frontiers and panel data, and time varying technical inefficiency. J Econ 46:201–211
Kumbhakar SC, Heshmati A (1995) Efficiency measurement in Swedish dairy farms: an application of rotating panel data, 1976–88. Am J Agric Econ 77:660–674
Kumbhakar S, Lovell K (2000) Stochastic frontier analysis. Cambridge University Press, Cambridge
Kumbhakar S, Ghosh S, McGuckin J (1991) A generalized production frontier approach for estimating determinants of inefficiency in U.S. Dairy farms. J Bus Econ Stat 9:279–286
Lee LF, Griffiths WE (1979) The prior likelihood and best linear unbiased prediction in stochastic coefficient linear models. Working Paper 1, University of New England
Lee YH, Schmidt R (1993) Production frontier model with flexible temporal variation in technicalefficiency. In: Fried HO, Lovell CAK, Schmidt SS (eds) The measurement of productive efficiency. Oxford University Press, New York, pp 237–255
Lee L, Tyler M (1978) The stochastic frontier production function and average efficiency. J Econ 7:385–390
Meeusen W, van den Broeck J (1977) Efficiency estimation from Cobb-Douglas production functions with composed error. Int Econ Rev 18:435–444
Mundlack Y (1961) Empirical production function free of management bias. J Farm Econ 43:44–56
Murillo-Zamoran LR, Vega-Cervera JA (2001) The use of parametric and nonparametric frontier methods to measure the productive efficiency in the industrial sector. A comparative study. Int J Product Econ 69(3):265–275
Nerlove M (1963) Returns to scale in electricity supply. In: Christ C et al (eds) Measurement in economics. Stanford University Press, Stanford
Nishimizu M, Page JM (1982) Total factor productivity growth, technological progress and technical efficiency change: dimensions of productivity change in Yugoslavia, 1967–1978. Econ J 92:920–936
Pitt M, Lee L (1981) The measurement and sources of technical inefficiency in the Indonesian weaving industry. J Dev Econ 9:43–64
Richmond J (1974) Estimating the efficiency of production. Int Econ Rev 15:515–521
Samuelson P (1938) Foundations of economic analysis. Harvard University Press, Cambridge

Schmidt P (1976) On the statistical estimation of parametric frontier production functions. Rev Econ Stat 58:238–239
Schmidt P, Sickles R (1984) Production frontiers with panel data. J Bus Econ Stat 2(4):367–374
Shephard R (1953) Cost and production functions. Princeton University Press, Princeton
Stevenson R (1980) Likelihood functions for generalized stochastic frontier functions. J Econ 13:57–66
Timmer P (1971) Using a probabilistic frontier production function to measure technical efficiency. J Polit Econ 79:776–794
Tsionas EG (2001) An introduction to efficiency measurement using Bayesian stochastic frontier models. Glob Bus Econ Rev 3(2):287–311
Tsionas EG (2002) Stochastic frontier models with random coefficients. J Appl Econ 17:127–147
Winsten CB (1957) Discussion on Mr. Farrell's paper. J R Stat Soc Ser A Stat Soc 120(3):282–284

Chapter 3
Introduction to Data Envelopment Analysis

Jorge Santos, Elsa Rosário Negas, and Luís Cavique Santos

Abstract This chapter introduces the basics of data envelopment analysis techniques, with a short historical introduction and examples of the constant returns to scale model (CRS) and the variable returns to scale (VRS) model. The ratio models are linearized and for both orientations primal and dual models are presented.

Keywords Data envelopment analysis • Efficiency evaluation

3.1 Introduction

Data envelopment analysis (DEA) is a mathematical programming-based technique to evaluate the relative performance of organisations. While the main applications have been in the evaluation of not-for-profit organisations, the technique can be successfully applied to other situations competing with other techniques as cost benefit analysis and multi criteria decision making as can be seen, for instance, in a recent study about the best choice for traffic planning, namely, the design and location of a highway in Memphis (Bougnol et al. 2005).

DEA is suited for this type of evaluation because it enables results to be compared making allowances for factors (Thanassoulis and Dunstan 1994). DEA makes it possible to identify efficient and inefficient units in a framework where results are

J. Santos (✉)
CIMA, Departamento de Matemática, Universidade de Évora, Colégio Luís Antº Verney
R. Romão Ramalho, 59 7000-671 Évora, Portugal
e-mail: jmas@uevora.pt

E.R. Negas
CITAD, Lusíada University of Lisbon, Portugal
e-mail: elsa.negas@lis.ulusiada.pt

L.C. Santos
Science and Technology Department, Universidade Aberta, Lisboa, Portugal
e-mail: lcavique@univ-ab.pt

A.B. Mendes et al. (eds.), *Efficiency Measures in the Agricultural Sector: With Applications*, DOI 10.1007/978-94-007-5739-4_3,

considered in their particular context. In addition, DEA also provides information that enables the comparison of each inefficient unit with its "peer group", that is, a group of efficient units that are identical with the units under analysis. These role-model units can then be studied in order to identify the success factors which other comparable units can attempt to follow. Thanassoulis (1993) argue that DEA is preferable to other methods, such as regression analysis, which also make it possible to contextualise results.

This chapter is structured as follows. The next section describes the development and fields of application of the technique, while Sect. 3.3 introduces the DEA models followed by a numerical and graphical example. Section 3.4 presents the mathematical formulation and the last one ends up with the main conclusions.

3.2 History and Applications of DEA

DEA is a mathematical programming technique presented in 1978 by Charnes et al. (1978), although its roots may be found as early as 1957 in Farrel's seminal work (Farrell 1957) or even to Debreu's, which introduced in the early fifties the "coefficient of resource utilisation" (Debreu 1951). It deserves special attention and also the work of the Dutch Nobel-prized Tjalling Koopmans and his "activity analysis concepts" (Koopmans 1951).

The DEA technique is usually introduced as a non-parametric one, but in fact it rests on the assumption of linearity (Chang and Guh 1991) and for the original constant returns to scale (CRS) models even in the more stringent assumption of proportionality.

Its application has been focused mainly on the efficiency assessment of not-for-profit organisations, since these cannot be evaluated on the basis of traditional economic and financial indicators used for commercial companies.

The first application of DEA was in the agriculture field; as a matter of fact, Farrell applied it to 1950 data of 48 states in the United States of America, considering 4 inputs and 2 outputs. At that time, the DEA term was not yet created, so in fact the first time the term DEA and that technique was applied was in the area of education, specifically in the analysis of Program Follow Through, conducted in the USA, in the late seventies (Rhodes 1978). Since then it has been used to assess efficiency in areas such as health (Wilson et al. 2012), county gaols (Seiford and Zhu 2002), courts (Schneider 2005), universities (Bougnol et al. 2010) and many other not-for-profit sectors. Nowadays DEA can be seen to have spread to other fields such as transit (Chiu et al. 2011), mining, (Chen et al. 2010), air transportation (Pestana e Dieke 2007) and even banking (Emrouznejad and Anouze 2010).

In data envelopment analysis, the organisational units to be assessed should be relatively homogeneous and were originally termed decision-making units (DMUs). As the whole technique is based on comparison of each DMU with all the remaining ones, a considerable large set of units is necessary for the assessment to be meaningful. We will assume that each DMU produces N outputs by means of M inputs.

3.3 The Meaning of DEA Efficiency

We will introduce some simple examples based on the following data set with 4 variables: 2 inputs – X_1 and X_2 – and 2 outputs – Y_1 and Y_2. Since this data will be used for pedagogical proposes, it is a small data set with just 12 decision-making units (DMU) (Table 3.1).

First, to introduce some basic concepts, we will suppose that only X_1 and Y_1 would be important for our analysis.

In this case, we can graph the 12 observations on a scatter plot, and it would be obvious that the most efficient one will be the DMU 3 since a straight line originating at the (0;0) point towards DMU 3 has the higher slope than any of the remaining (Fig. 3.1).

The straight line originating at the (0;0) point towards DMU 3 is called the efficiency frontier and together with the X axis it defines a cone with its vertex

Table 3.1 Illustrative data set

DMU	X_1	X_2	Y_1	Y_2
1	4,0	140,0	2,0	28,0
2	5,0	90,0	1,0	22,5
3	6,0	36,0	6,0	12,0
4	10,0	300,0	8,0	60,0
5	11,0	66,0	7,0	16,5
6	8,0	36,0	6,0	12,0
7	9,0	12,0	7,0	6,0
8	5,0	210,0	3,0	30,0
9	5,5	33,0	4,4	5,5
10	8,0	288,0	4,0	72,0
11	10,0	80,0	2,0	20,0
12	8,0	8,0	1,0	4,0

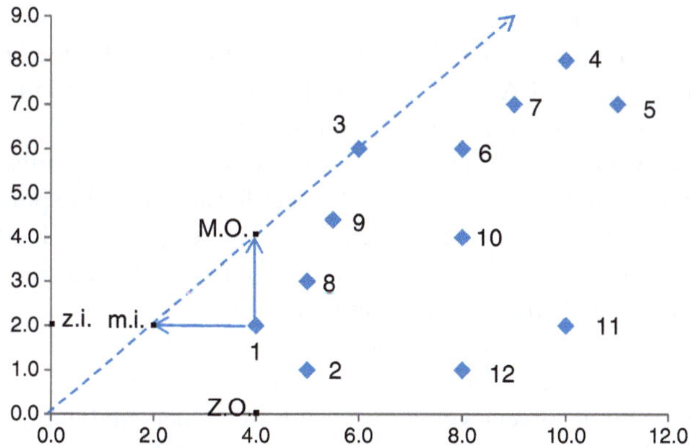

Fig. 3.1 Example with 12 hypothetical farms consuming X_1 and producing Y_1

C:\com\0jj\d\0UE\Springer\DEA\X1Y1.xls_CRS_RAD_IN

	DMU	Score	X1{I}{V}	Y1{O}{V}	Benchmarks	{S} X1{I}	{S} Y1{O}
1	Farm 1	50,00%	1,00000	1,00000	3 (0,33)	0,00000	0,00000
2	Farm 2	20,00%	1,00000	1,00000	3 (0,17)	0,00000	0,00000
3	Farm 3	100,00%	1,00000	1,00000	11		
4	Farm 4	80,00%	1,00000	1,00000	3 (1,33)	0,00000	0,00000
5	Farm 5	63,64%	1,00000	1,00000	3 (1,17)	0,00000	0,00000
6	Farm 6	75,00%	1,00000	1,00000	3 (1,00)	0,00000	0,00000
7	Farm 7	77,78%	1,00000	1,00000	3 (1,17)	0,00000	0,00000
8	Farm 8	60,00%	1,00000	1,00000	3 (0,50)	0,00000	0,00000
9	Farm 9	80,00%	1,00000	1,00000	3 (0,73)	0,00000	0,00000
10	Farm 10	50,00%	1,00000	1,00000	3 (0,67)	0,00000	0,00000
11	Farm 11	20,00%	1,00000	1,00000	3 (0,33)	0,00000	0,00000
12	Farm 12	12,50%	1,00000	1,00000	3 (0,17)	0,00000	0,00000

Fig. 3.2 Example of EMS results under constant returns to scale minimisation of inputs

at the origin. This cone is called the production possibility set, since it contains all real data, and according to DEA axioms, only points inside this cone correspond to possible working conditions based on best-achieved performance.

We will analyse in greater detail unit 1. There are two ways for DMU 1 to reach efficiency: increasing output, till it reaches M.O. (maximisation of output) or reducing the input till m.i. (minimisation of input).

The actual value for efficiency is defined as the ratio between the distances: $d\,(\text{m.i.} - \text{z.i.})\,/d\,(1 - \text{z.i.}) = (2 - 0)\,/(4 - 0) = 50\%$. On a similar way, the actual value for inefficiency is defined as the ratio between the distances: $d\,(\text{M.O.} - \text{Z.O.})\,/d\,(1 - \text{Z.O.}) = (4 - 0)\,/(2 - 0) = 100\%$. In these calculations, z.i. means the zero input point and Z.O. the zero output point for the DMU 1.

We conclude that in DEA, there are two different means of optimization related with radial measures of score: input minimisation and output maximisation. The artificial points M.O. and m.i. are termed targets or composite points for DMU 1. Point 3 is the only efficient one, and it is the peer for all remaining points.

Under constant returns to scale, efficiency is the reciprocal of inefficiency; the peer set is also the same, regardless of the orientation, although their targets are different as we conclude from the exposition above.

Typical DEA software like efficiency measurement system (EMS), developed by Holger Scheel (Scheel 2000) at Dortmunt University (using Csaba Mészáros' BPMPD interior-point solver) would give us results as depicted in Fig. 3.2.

EMS highlights the efficient units and provides us all the necessary values:

- *Score* means efficiency.
- $X_1\{I\}\{V\}$ presents the virtual input of variable X_1, in a similar way $Y_1\{O\}\{V\}$ presents the virtual output of variable Y_1.

C:\com\0jj\d\0UE\Springer\DEA\X1Y1.xls_CRS_RAD_OUT

	DMU	Score	X1{I}{V}	Y1{O}{V}	Benchmarks	{S} X1{I}	{S} Y1{O}
1	Farm 1	200,00%	1,00000	1,00000	3 (0,67)	0,00000	0,00000
2	Farm 2	500,00%	1,00000	1,00000	3 (0,83)	0,00000	0,00000
3	Farm 3	100,00%	1,00000	1,00000	11		
4	Farm 4	125,00%	1,00000	1,00000	3 (1,67)	0,00000	0,00000
5	Farm 5	157,14%	1,00000	1,00000	3 (1,83)	0,00000	0,00000
6	Farm 6	133,33%	1,00000	1,00000	3 (1,33)	0,00000	0,00000
7	Farm 7	128,57%	1,00000	1,00000	3 (1,50)	0,00000	0,00000
8	Farm 8	166,67%	1,00000	1,00000	3 (0,83)	0,00000	0,00000
9	Farm 9	125,00%	1,00000	1,00000	3 (0,92)	0,00000	0,00000
10	Farm 10	200,00%	1,00000	1,00000	3 (1,33)	0,00000	0,00000
11	Farm 11	500,00%	1,00000	1,00000	3 (1,67)	0,00000	0,00000
12	Farm 12	800,00%	1,00000	1,00000	3 (1,33)	0,00000	0,00000

Fig. 3.3 Example of EMS results under constant returns to scale maximisation of outputs

- *Benchmarks* means that all farms, except for the third, are benchmarked against farm 3 and so is used 11 times for comparison. The coefficients in this column mean that if we multiply farm 3 by those values, it will result in the composite-projected target point.
- $\{S\}X_1\{I\}$ presents the slacks for input X_1, in a similar way $\{S\}Y_1\{O\}$ presents the slacks for output Y_1.

Thinking again around unit 1, it means that its target under input minimisation is $0{,}33 \times (6;6) = (2;2)$. These are exactly the coordinates of point m.i. in Fig. 3.1. The concepts of virtual multipliers and slacks will be clarified after the presentation of the mathematical formulation.

So far it is very important to emphasise that there are two different radial orientations: input minimisation and output maximisation. There are target points and a set of efficient units that benchmark the non-efficient ones.

The results for output maximisation are depicted on Fig. 3.3.

We can confirm that the conclusions are very similar to those from Fig. 3.2; more specifically the score now is the inefficiency, the reciprocal of the scores under input minimisation. Slacks are identically null as in the minimisation of inputs.

We can proceed with our analysis including output Y_2 in our analysis. In this case we will have 3 variables, and we cannot represent them on the *XY* plane. Anyway, since we assume that constant returns to scale prevail, we can normalise the outputs by the single input X_1. That way, we get the Fig. 3.4, where we can easily spot three efficient farms: Farm 3 remains efficient (whenever we add new variables, efficiencies never decrease), and farms 4 and 10 turn out efficient, joining the new efficiency frontier.

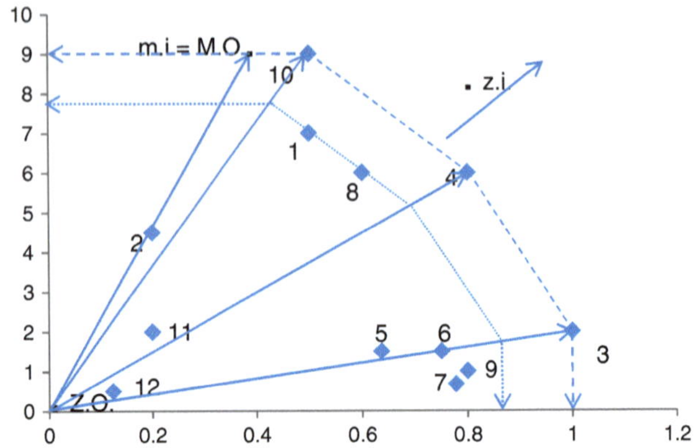

Fig. 3.4 Cut of the PPS at $X_1 = 1$, under CRS, two outputs: Y_1 and Y_2

We point out in Fig. 3.4 the zero output point (Z.O.) at the origin and the zero input area that lies outside the production possibility space (PPS). The PPS is now a polyhedral cone (an inverted pyramid) with 3 specific edges: the rays that depart from the origin to each of the 3 efficient units (3;4;10). There are other 3 edges that are not specific, since they exist always regardless of the specific data set; one belongs to the plane defined by $X = 0$, the other belongs to the plane with $Y = 0$ and finally the third is the Z axis with both $X = 0$ and $Y = 0$ (Fig. 3.4).

The graph in Fig. 3.4 is a cut of the PPS at $X_1 = 1$, since we normalised the outputs. The dashed line is the intersection of the horizontal plane $X_1 = 1$ with the efficiency frontier (the polyhedral cone). The dotted line is an isoefficiency contour.

It is clear that DEA clusters DMU according to their specificities; farms 1 and 8 are related to farms 4 and 10, their peer group. The same way farm 6 will be similar to farm 3, and so on. A different situation arises with farm 2, which cannot be expressed as a finite linear positive combination of the efficient ones; in that case, it is necessary to have a slack in Y_1. The same situation happens with farms 7 and 9, they "need" slacks in Y_2, and this means that after their projection on the vertical part of the dashed line, they still have to increase its production of Y_2. This can be confirmed by the EMS results for this case. Typically this kind of figure is associated with input minimisation, since we have only one input to reduce. It can be associated also to output maximisation, where an equiproportional maximisation of both outputs (keeping the output mix constant) is required.

Finally, we examine the case for two inputs and a single output Y_2. This time we cannot anticipate the results, since we did not study efficiencies for pairs of these three variables. Again, since we assume that constant returns to scale prevail, we can normalise the inputs by the single output Y_2. That way we get Fig. 3.5 where we plotted X_2/Y_2 versus X_1/Y_2 where we can easily spot three efficient farms:

The PPS is again a polyhedral cone (an inverted unbounded pyramid) with 3 specific edges: the rays that depart from the origin to each of the 3 efficient units (3;7;10). There are no other edges since the inefficient units are located in an unbounded region that tends to infinity as outputs approach zero.

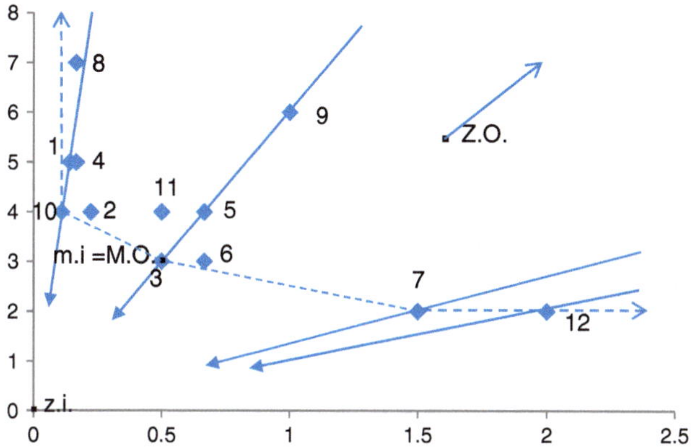

Fig. 3.5 Cut of the PPS at $Y_2 = 1$, under CRS, two inputs: X_1 and X_2

	DMU	Score	X1{I}{V}	X2{I}{V}	Y2{O}{V}	Benchmarks	{S} X1{I}	{S} X2{I}	{S} Y2{O}
1	Farm 1	125,24%	0,06844	0,93156	1,00000	3 (0,02) 10 (0,48)	0,00000	0,00000	0,00000
2	Farm 2	106,67%	0,12500	0,87500	1,00000	3 (0,50) 10 (0,25)	0,00000	0,00000	0,00000
3	Farm 3	100,00%	0,23077	0,76923	1,00000	7			
4	Farm 4	126,67%	0,07895	0,92105	1,00000	3 (0,33) 10 (1,00)	0,00000	0,00000	0,00000
5	Farm 5	133,33%	0,21751	0,78249	1,00000	3 (1,83)	0,00000	0,00000	0,00000
6	Farm 6	104,76%	0,18182	0,81818	1,00000	3 (0,90) 7 (0,29)	0,00000	0,00000	0,00000
7	Farm 7	100,00%	0,33333	0,66667	1,00000	2			
8	Farm 8	150,00%	1,00000	0,00000	1,00000	10 (0,62)	0,00000	30,00000	0,00000
9	Farm 9	200,00%	0,22861	0,77139	1,00000	3 (0,92)	0,00000	0,00000	0,00000
10	Farm 10	100,00%	1,00000	0,00000	1,00000	5			
11	Farm 11	123,33%	0,24324	0,75676	1,00000	3 (1,56) 10 (0,08)	0,00000	0,00000	0,00000
12	Farm 12	100,00%	0,00000	1,00000	1,00000	7 (0,67)	2,00000	0,00000	0,00000

Fig. 3.6 EMS results for the two inputs, X_1 and X_2, and a single output, Y_2

Farm 12 is not strongly efficient, since farm 7 has the same ratio on X_2/Y_2, but a smaller one on X_1/Y_2. Indeed there is a slack on X_1. We can see in the graph that the two points differ 0.5 on X_1/Y_2. This happens at the cutting plane $Y_2 = 1$, as the true scale of operation of farm 12 is Y_2. The slack is $0.5 \times 4 = 2$ as can be seen in Fig. 3.6 where we present the results for this case. We can notice that the only farms that EMS classifies as efficient are farms 3, 7 and 10, those that are on shadowed lines of the table. Farm 12 presents a score of 100%, although it is not efficient since it has 2-unit slack in X_1; its efficiency according to the Charnes Cooper and Rhodes model (CCR model) is $100\% - 2\varepsilon$ where ε is a small non-Archimedean entity. These units that are on the "horizontal or vertical" parts of the efficiency frontier are termed "weakly efficient DMUs"; we will come back to this matter when we present the models in the remaining of this section.

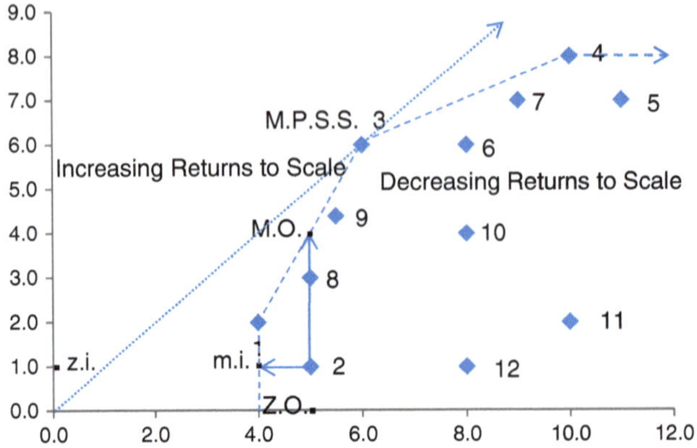

Fig. 3.7 Comparison of efficiency frontiers of the CCR and BCC models

So far we have assumed we are operating in a constant returns to scale situation. We will now present only one graphical example for the variable returns to scale (VRS) model, introduced by Banker, Charnes and Cooper in 1984 (Banker et al. 1984).

We will study again the one input (X_1) producing one single output (Y_1) case, but according to Banker, Charnes and Cooper the efficiency frontier is not anymore a polyhedral cone, but instead the intersection of all the polyhedral cones, one for each DMU; this result is based on the convex hull of the all DMUs. In the CCR model, we had flexibility on the choice of the weights; now in the BCC model, we are free to choose also the origin of our data (the vertex of the polyhedral cone).

In Fig. 3.7, we have a vertical facet that ends at DMU 1; this facet means that we assume a fixed amount of input ($X_1 = 4$) necessary just to start the business. DMU 2 under VRS will have a much higher efficiency than under CRS, especially in the minimisation of inputs case. DMU 2 is spending 5 units of input, and it was expected to spend just one in the CRS case; this leads to a CRS efficiency of 1/5 = 20%; in the more benevolent BCC model, it was expected to spend 4 units, so its efficiency rises to 4/5 = 80%. The ratio between CRS efficiency and VRS efficiency is called scale efficiency and increases from the origin (X=0; Y=0) till DMU 3 and then starts decreasing. The first region is the increasing returns to scale region, and the second is the decreasing returns to scale region. In higher dimensions it seems more difficult, but it is much easier than it seems at first look. The most productive scale size (MPSS) is simply the intersection between the VRS and the CRS efficiency frontiers. From the origin till the MPSS, we are in the increasing returns to scale region; from the MPSS till infinity, we are on the decreasing returns to scale region.

In CRS, we had a simple relation between the scores of input minimisation and output maximisation (they were reciprocals: efficiency = 1/inefficiency). Under VRS it does not happen anymore. In fact there is no clear relation between those two scores since the peer set most of the times differs from one orientation to the other. For instance, DMU 10 under input minimisation has DMU 1 and 3 as peers, but if we intend to maximise its output, the reference set is 3 and 4.

C:\com\0jj\d\0UE\Springer\DEA\X1Y1.xls_VRS_RAD_IN

	DMU	Score	X1{I}{V}	Y1{O}{V}	Benchmarks	{S} X1{I}	{S} Y1{O}
1	Farm 1	100,00%	1,00000	1,00000	6		
2	Farm 2	80,00%	1,00000	1,00000	1 (1,00)	0,00000	1,00000
3	Farm 3	100,00%	1,00000	1,00000	6		
4	Farm 4	100,00%	1,00000	1,00000	2		
5	Farm 5	72,73%	1,00000	1,00000	3 (0,50) 4 (0,50)	0,00000	0,00000
6	Farm 6	75,00%	1,00000	1,00000	3 (1,00)	0,00000	0,00000
7	Farm 7	88,89%	1,00000	1,00000	3 (0,50) 4 (0,50)	0,00000	0,00000
8	Farm 8	90,00%	1,00000	1,00000	1 (0,75) 3 (0,25)	0,00000	0,00000
9	Farm 9	94,55%	1,00000	1,00000	1 (0,40) 3 (0,60)	0,00000	0,00000
10	Farm 10	62,50%	1,00000	1,00000	1 (0,50) 3 (0,50)	0,00000	0,00000
11	Farm 11	40,00%	1,00000	1,00000	1 (1,00)	0,00000	0,00000
12	Farm 12	50,00%	1,00000	1,00000	1 (1,00)	0,00000	1,00000

Fig. 3.8 EMS results for the one input, X_1, and a single output, Y_1, under VRS

In Fig. 3.8, we present the results for the one input, X_1, and a single output, Y_1, under VRS and minimisation of inputs. We cannot represent graphically any 3 variables case under variable returns to scale, so now we have to introduce the mathematical formulation that relies on linear programming and duality.

3.4 Mathematical Formulation for DEA

In DEA, efficiency (Ef_a) of a specific decision-making unit (DMU_a) under analysis is defined as the ratio between a weighted sum of its s outputs Y_{ra} and a weighted sum of its m inputs X_{ia}, a natural extension of the concept of efficiency used in the fields of physics and engineering (Charnes et al. 1978).

$$\mathrm{Ef}_a = \frac{\sum_{r=1}^{s} \mu_{ra} y_{ra}}{\sum_{i=1}^{m} v_{ia} x_{ia}} \tag{3.1}$$

When assessing a set of J organisations, where X_{ik} stands for the ith input of the kth DMU, with a similar meaning for Y_{rk}, the weights μ_{rk} and v_{ik}, in Eq. (3.1), are chosen for each DMU_j, under evaluation as those that maximise its efficiency as defined by Ef_a. Several constraints have to be added to the maximisation problem:

- The strict positivity (Charnes et al. 1978) of the weights μ_{rk} and v_{ik}, (also known as virtual multipliers).

- For scaling purposes, all n DMUs under analysis must have efficiencies not exceeding an agreed value, typically one or 100%, as is usual in engineering definitions of efficiency.
- A third kind of restriction has to be included since otherwise this linear fractional program would yield an infinite number of solutions. In fact, if a set of weights μ_{rk} and v_{ik} returns the optimal solution, so would $k\mu_{rk}$ and kv_{ik}. Making the denominator in Eq. (3.1) equal to one or 100% circumvents this situation.

So, we have to solve the following linear programmingmaximisation problem *for each one* of the J DMUs under analysis:

$$\max \ \mathrm{Ef}_a = \frac{\sum_{r=1}^{s} \mu_{ra} y_{ra}}{\sum_{i=1}^{m} v_{ia} x_{ia}} \tag{3.2}$$

$$\text{s.t.} \ \ \mu_{ra} \geq \varepsilon > 0 \quad r = 1 \ldots s \tag{3.3}$$

$$v_{ia} \geq \varepsilon > 0 \quad i = 1 \ldots m \tag{3.4}$$

$$\mathrm{Ef}_k = \frac{\sum_{r=1}^{s} \mu_{rk} y_{rk}}{\sum_{i=1}^{m} v_{ik} x_{ik}} \leq 1 \quad k = 1 \ldots n \tag{3.5}$$

$$\sum_{i=1}^{m} v_{ia} x_{ia} = 1 \tag{3.6}$$

This fractional linear program can be solved by the Charnes and Cooper transformation (Charnes and Cooper 1962) which yield the following linear program:

$$\text{Max } \mathrm{Ef}_a = \sum_{r=1}^{s} \mu_{ra} y_{ra} \tag{3.7}$$

s.t.

$$\sum_{i=1}^{m} v_{ia} x_{ia} = 1 \tag{3.8}$$

$$\sum_{i=1}^{m} v_{ia} x_{ik} - \sum_{r=1}^{s} \mu_{ra} y_{rk} \leq 0 \quad j = 1 \ldots J \tag{3.9}$$

$$\mu_{ra} \geq \varepsilon > 0 \quad r = 1 \ldots s \tag{3.10}$$

$$v_{ia} \geq \varepsilon > 0 \quad i = 1 \ldots m \tag{3.11}$$

The problem above is known as the multiplier problem, since its unknowns are the weights, which are usually lower bounded by a small quantity non-Archimedean entity, ε (Eqs. 3.10 and 3.11), so that all inputs and outputs are considered in the evaluation (Charnes et al. 1979) even if with a minor weight ε, set typically equal to 10^{-6}.

The dual of this problem, which we shall call the envelopment problem, provides important information about economies that could be achieved in all the inputs; it also indicates which efficient units the inefficient unit being assessed should emulate. Those efficient units are usually referred to as the reference set or peer group of the unit under evaluation.

Any linear problem has a dual, which we will call the envelopment problem that can be written as follows:

$$\min : Z_{\mathrm{a}} - \varepsilon \left\{ \sum_{r=1}^{s} S_{ra}^{+} + \sum_{i=1}^{m} S_{ia}^{-} \right\} \tag{3.12}$$

$$\text{s.t.} \sum_{k=1}^{n} \lambda_k x_{ik} + S_{ia}^{-} = Z_a x_{ia} \qquad i = 1 \ldots m \tag{3.13}$$

$$\sum_{k=1}^{n} \lambda_k y_{rk} - S_{ra}^{+} = y_{ra} \qquad r = 1 \ldots s \tag{3.14}$$

$$\lambda_k \geq 0 \qquad k = 1 \ldots n \tag{3.15}$$

$$S_{ra}^{+} \geq 0 \qquad r = 1 \ldots s \tag{3.16}$$

$$S_{ia}^{-} \geq 0 \qquad i = 1 \ldots m \tag{3.17}$$

This formulation can be interpreted as follows: Given $\text{DMU}_{a,}$ find the composite unit which has no smaller outputs than this one and whose inputs are smaller than those of DMU_a scaled down by a factor Z_a as small as possible. This is why this formulation is known as input minimisation; since we are minimising Z_a, we are seeking the minimal inputs that, based on best-achieved performance, could still produce the same amount of outputs as DMU_a is currently producing. These composite units are finite linear combinations of efficient units more specifically finite positive combinations which lie on the efficiency frontier. These efficient DMUs are known as the *peer group*, the role-model units that inefficient DMU_a should try to emulate.

In the envelopment problem, it is easy to understand the role of the small non-Archimedean entities ε; they simply multiply by the sum of slacks so that slacks are not ignored in the overall score.

As a whole, the interpretation of the DEA technique is straightforward and can be put in the following terms:

3.4.1 Multiplier Problem

Evaluate each DMU with the set of weights which maximises its efficiency, provided that all other DMUs, rated with that set of weights, have efficiency not greater than unity.

3.4.2 Envelopment Problem

Find the smallest proportion Z_a of inputs that would bring the current DMU_a to the enveloping surface of all DMUs.

The model presented above, named CCR after the acronym of the authors Charnes et al. (1978), assumes constant returns to scale. This means that when the input of an efficient unit is multiplied by a given factor, its output level is also multiplied by the same factor. In this case, the production possibilities set will be a closed convex polyhedral cone in $\mathcal{R}^{n+m}$ in the positive orthant.

In other situations this is not the case. The scale of operations may have an impact on the outputs, creating "economies" or "diseconomies" of scale. The BCC model, developed by Banker et al. (1984), can deal with variable returns to scale.

The BCC envelopment model can be obtained from the CCR envelopment model by adding the convexity constraint to the envelopment problem:

$$\sum_{k=1}^{n} \lambda_k = 1 \tag{3.18}$$

The multiplier problem is the dual of the envelopment formulation, and the extra restriction originates an extra free variable. It is important to note that if we relax the convexity constraint to $\sum_{k=1}^{n} \lambda_k \geq 1$ than its dual, variable will become non-negative and it becomes clear that this model (the non-decreasing returns to scale model) is nothing else but a CCR model with an extra output identically unitary.

The same analysis could be conducted based on the maximisation of outputs, leading to similar formulations.

For instance, DMU 10 will be rated under VRS as 5/8 efficient and 4/8 = 50% under CRS. It is straightforward to conclude that VRS efficiency is never smaller than the CRS efficiency. This became clear if we note that the envelopment problem has an extra constraint, and it is a minimising program. The ratio between those two efficiencies is called scale efficiency, and it is important in the efficiency analysis. All those definitions can be made also for the output maximisation orientation.

In VRS we can define several regions, the increasing returns to scale and the decreasing returns to scale that expands until the infinity. Those regions are separated by the most productive scale size, where the frontier of the pointed cone intersects the convex hull of the DMUs. In Fig. 3.7 it corresponds to point 3, the intersection between the dotted line (a linear hyperplane of dimension 1) and the dashed polygon.

3.5 Concluding Remarks

Using all the typical graphical representations of data envelopment analysis, the authors introduced based on a visual and spatial approach the basic ideas and principles of DEA. The two models for CRS and VRS, known as the CCR and BCC models, respectively, were presented, as well as both input minimisation and output maximisation orientations; this makes a total of 4 different models, and for each of those 4, there is a dual (weights = multipliers versus envelopment problems).

Many new developments of DEA could be presented here, most specifically weights restrictions, non-controllable inputs and outputs and what may be the most important one, the superefficiency model, that was introduced by Andersen and Petersen from Odense University by 1993 (Andersen and Petersen 1993). This technique is rather appealing because they allow efficiency to be greater than one, discriminating between efficient DMUs that otherwise are all ranked equal; it avoids also ambiguity on weights of those units (multiple solutions on the weights = multipliers problem, degeneracy in its dual).

The equations presented allow the implementation of DEA software in many programs as Excel, SAS, Mosel, Stata, OpenOffice and on any other program with linear programming solving capabilities.

References

Andersen P, Petersen NC (1993) A procedure for ranking efficient units in data envelopment analysis. Manag Sci 39(10):1261–1264

Banker RD, Charnes A, Cooper WW (1984) Some models for estimating technical and scale inefficiencies in data envelopment analysis. Manag Sci 30(9):1078–1092

Bougnol M-L, Dulá JH, Retzlaff-Roberts D, Womer NK (2005) Nonparametric frontier analysis with multiple constituencies. J Oper Res Soc 56(3):252–266

Bougnol M-L, Dulá JH, EstellitaLins MP, Moreira da Silva AC (2010) Enhancing standard performance practices with DEA. Omega 38(1–2):33–45. doi:10.1016/j.omega.2009.02.002

Chang K-P, Guh Y-Y (1991) Linear production functions and the data envelopment analysis. Eur J Oper Res 52(2):215–223

Charnes A, Cooper WW (1962) Programming with linear fractional functionals. Nav Res Logist Q 9:181–185

Charnes A, Cooper WW, Rhodes E (1978) Measuring the efficiency of decision making units. Eur J Oper Res 2(6):429–444

Charnes A, Cooper WW, Rhodes E (1979) Short communication: measuring the efficiency of decision making units. Eur J Oper Res 3(4):339

Chen X, Chen X, Zhou H, Zhou H (2010) Coal enterprises safety investment efficiency analysis based on data envelopment. WKDD '10 Proceedings of the 2010 third international conference on knowledge discovery and data mining. IEEE Computer Society, Washington, DC, USA, pp 175–178, ©2010, ISBN: 978-0-7695-3923-2, doi: 10.1109/WKDD.2010.73

Chiu Y-h, Huang C-w, Ma C-M (2011) Assessment of China transit and economic efficiencies in a modified value-chains DEA model. Eur J Oper Res 209(2):95–103

Debreu G (1951) The coefficient of resource utilization. Econometrica 19:273–292

Emrouznejad A, Anouze AL (2010) Data envelopment analysis with classification and regression tree-a case of banking efficiency. Expert Syst 27(4):231–246

Farrell MJ (1957) The measurement of productive efficiency. J R Stat Soc A 120:253–281

Koopmans T (1951) Analysis of production as an efficient combination of activities. In: Koopmans TC (ed) Activity analysis of production and allocation. Wiley, New York

Pestana Barros C, Dieke PUC (2007) Performance evaluation of Italian airports: a data envelopment analysis. J Air Transp Manag 13(4):184–191

Rhodes E (1978) Data envelopment analysis and related approaches for measuring the efficiency of decision-making units with an application to program follow through in U.S. education. Unpublished doctoral thesis, School of Urban and Public Affairs, Carnegie – Mellon University

Scheel H (2000) EMS: efficiency measurement system user's manual version 1.3. http://www.microtheory.uni-jena.de/download/ems.pdf. Accessed 21 Sept 2012

Schneider M (2005) Judicial career incentives and court performance: an empirical study of the German labour courts of appeal. Eur J Law Econ 20(2):127–144

Seiford LM, Zhu J (2002) Value judgment versus allocative efficiency: a case of Tennessee county jails. J Manag Sci Reg Dev 4(July):89–98

Thanassoulis E (1993) Comparison of regression analysis and data envelopment analysis as alternative methods for performance assessments. J Oper Res Soc 44(11):1129–1144

Thanassoulis E, Dunstan P (1994) Guiding schools to improved performance using data envelopment analysis: an illustration with data from a local education authority. J Oper Res Soc 45(11):1247–1262

Wilson Asa B, Kerr Bernard J, Bastian Nathaniel D, Fulton Lawrence V (2012) Financial performance monitoring of the technical efficiency of critical access hospitals: a data envelopment analysis and logistic regression modeling approach. (Report). J Healthc Manag 57(3):200–212; discussion 212–3

Chapter 4
Superefficiency and Multiplier Adjustment in Data Envelopment Analysis

Jorge Santos, Luís Cavique Santos, and Armando B. Mendes

Abstract Superefficiency is an important extension of DEA that overcomes some limitations of the traditional models, specifically allowing ranking of efficient units and a unique set of weights for those units. Weights restriction is a well-known technique in the DEA field. When those techniques are applied, weights cluster around its new limits, making its evaluation dependent of its levels. This chapter introduces a new approach to weights adjustment by goal programming techniques, avoiding the imposition of hard restrictions that can even lead to unfeasibility. This method results in models that are more flexible.

Keywords Data envelopment analysis • Superefficiency • Weights restriction • Evaluation • Goal programming

4.1 Introduction

Data envelopment analysis (DEA) is a linear programming-based technique to assess the relative performance of organisations. While the main applications have been in the evaluation of not-for-profit organisations, the technique has been successfully applied to other organisations too.

J. Santos (✉)
CIMA, Departamento de Matemática, Universidade de Évora, Colégio Luís Antº Verney
R. Romão Ramalho, 59 7000-671 Évora, Portugal
e-mail: jmas@uevora.pt

L.C. Santos
LabMAg, Science and Technology Department, Universidade Aberta, Lisboa, Portugal
e-mail: lcavique@univ-ab.pt

A.B. Mendes
CEEAplA, Departamento de Matemática, Universidade dos Açores, Rua da Mãe de Deus, 9500-801 Ponta Delgada, Açores, Portugal
e-mail: amendes@uac.pt

A.B. Mendes et al. (eds.), *Efficiency Measures in the Agricultural Sector: With Applications*, DOI 10.1007/978-94-007-5739-4_4,

With this chapter, we have two objectives in mind. The first one is to present superefficiency in data envelopment analysis (DEA), an extension to the technique which may have useful applications in many evaluation contexts, namely, when we have many efficient DMUs with ambiguous weights and we want to rank those 100% efficient DMUs. In addition to allowing the ranking of these organisations traditionally classified with a score of 100%, superefficiency also known as the deleted domain technique also creates the conditions to improve performance through target setting and role model identification by means of a unique set of weights and a nondegenerate dual linear program.

The second objective is to introduce an entirely new way of adjusting multipliers by means of goal programming techniques. This adjustment is a much more general way of dealing with the incorporation of exogenous structure preferences so far relying only in weights restriction techniques, which, in our point of view, leads to the concentration of the weights in its upper and lower limits.

This chapter is structured as follows. The next introduces the DEA models followed by a numerical example useful for introducing the superefficiency evaluation, an extension of DEA also known as deleted domain. Sections 4.5 and 4.6 deal with the graphical solution in the weights space and make a very short description of the weights restrictions technique, respectively.

In Sect. 4.7, a new concept of multiplier adjustment is introduced and exemplified through a small data set. In Sect. 4.6, a case with artificially generated data is solved to highlight the potentialities of this technique. This chapter ends up with a final section with the conclusions and directions of future work. Readership, who wish to follow on DEA subject, should consider consulting the good review in Boussofiane et al. (1991).

4.2 Superefficiency/Deleted Domain Extension

When assessing a set of j organisations, where X_{mj} stands for the mth input of the jth DMU, with a similar meaning for Y_{nj}, the weights $\mu_{mj'}$ and $v_{nj'}$, in Eqs. (4.1), (4.2), (4.3), (4.4), and (4.5), are chosen for each j' DMU under evaluation as those that maximise its efficiency as defined by $h_{j'}$:

$$\text{Max } h_{j'} = \sum_{n=1}^{N} v_{nj'} y_{nj'} \tag{4.1}$$

$$\text{s.t.} \sum_{m=1}^{M} \mu_{mj'} x_{mj'} = 1 \tag{4.2}$$

$$\sum_{n=1}^{N} v_{nj'} y_{nj} \leq \sum_{m=1}^{M} \mu_{mj'} x_{mj} \quad j = 1 \ldots J \tag{4.3}$$

Table 4.1 Outputs normalised by input X_1

Farm	x_1	y_1	y_2	y_1/x_1	y_2/x_1
1	4.0	2.0	28.0	.500	7.000
2	5.0	1.0	22.5	.200	4.500
3	6.0	6.0	12.0	1.000	2.000
4	10.0	8.0	60.0	.800	6.000
5	11.0	7.0	16.5	.636	1.500
6	8.0	6.0	12.0	.750	1.500
7	9.0	7.0	6.0	.778	.667
8	5.0	3.0	30.0	.600	6.000
9	5.5	4.4	5.5	.800	1.000
10	8.0	4.0	72.0	.500	9.000
11	10.0	2.0	20.0	.200	2.000
12	8.0	1.0	4.0	.125	.500

$$\mu_{mj'} \geq \varepsilon > 0 \quad m = 1 \ldots M \tag{4.4}$$

$$\nu_{nj'} \geq \varepsilon > 0 \quad n = 1 \ldots N \tag{4.5}$$

The problem above is known as the multiplier problem, since its unknowns are the multipliers also known as weights; the dual of this problem is the envelopment problem that provides important information, indicating which efficient units the inefficient unit being assessed should emulate. Those efficient units are usually referred to as the reference set or peer group of the unit under evaluation (Charnes et al. 1978).

To illustrate the deleted domain data envelopment analysis technique, also known as superefficiency, an example is introduced in Table 4.1, with 12 DMUs producing two outputs Y_1 and Y_2 from a single input X_1, under the assumption of constant returns to scale (CRS), which simply means that if one doubles the inputs of any unit, it would be expected that its outputs would also double. In algebraic form, this can be stated as follows: if x_j yields outputs y_j then inputs kx_j should produce outputs ky_j.

In this simple example, we can normalise the outputs by the only input and plot them in the plane. This is also equivalent to consider that we are dealing with a constant input of 1. This way we will be working in the plane defined by $x = 1$ in the 3-dimensional one-input two-output space.

We arrive at the concept of superefficiency by allowing the efficiency of the DMU being assessed to be greater than unity. This is achieved by removing the corresponding constraint from the set of j constraints in Expression (4.3). This is the reason why this technique is also known as deleted domain. The superefficiency only affects units deemed as efficient, as the removed constraint is not binding for the inefficient units, since their efficiency is, by definition, less than unity.

The extension of this technique to the variable returns to scale(VRS) case (Banker et al. 1984) is easy. It suffices to add the convexity constraint to the envelopment problem, although it brings unfeasibility problems with some efficient DMUs (Santos 2008).

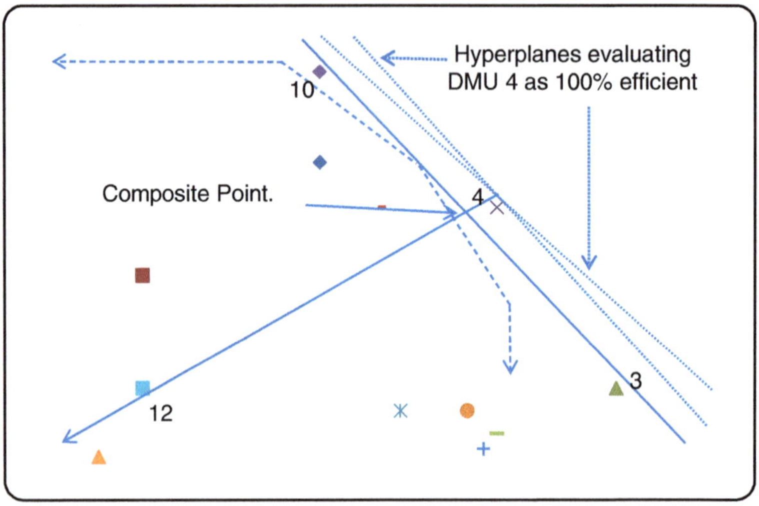

Fig. 4.1 Efficiency frontier and hyperplanes evaluating farm 4 as 100% and even 107.50% efficient

Table 4.2 Superefficiency scores for the efficient units

Unit	Superefficiency (%)
Unit 4	107.50
Unit 3	125.00
Unit 10	128.57

The basic idea is depicted in Fig. 4.1 where we represent two hyperplanes (the dotted lines crossing at DMU 4) assessing DMU 4 as 100% efficient (there is an infinity of those hyperplanes, since the multiplier problem in DEA has multiple solutions and is degenerate in its dual). If we allow the efficiency of DMU 4 to be greater than unity, then DMUs 3 and 10 become 100% efficient when assessed by the new hyperplane (the solid line in Fig. 4.1), and the superefficiency will be measured by the ratio between the distances to the origin of the composite point and the DMU 4. This new hyperplane is unique and the solution for the envelopment problem is not anymore degenerate, but it consists of DMUs 3 and 10.

This extension to DEA was first published by Andersen and Petersen from Odense University in 1993 (Andersen and Petersen 1993), and its use is strongly recommended by the authors as a consequence of its simplicity and usefulness.

By using superefficiency, it is possible to rank all units, even the efficient ones that by standard DEA techniques would all be rated as equal – their efficiency having reached the top value of 100%.

For the example presented in this section, the superefficiency for the 3 efficient DMUs would be as presented in Table 4.2.

Units 3 and 10 are efficient and robust, while any small increase in the input or decrease in the outputs of unit 4 may make it inefficient.

C:\com\0jj\d\0UE\Springer\DEA\X1Y1Y2.xls_CRS_RAD_IN

	DMU	Score	X1{I}{V}	Y1{O}{V}	Y2{O}{V}	Benchmarks	{S} X1{I}	{S} Y1{O}	{S} Y2{O}
1	Farm 1	85.71%	1.00000	0.35714	0.50000	4 (0.10) 10	0.00000	0.00000	0.00000
2	Farm 2	50.00%	1.00000	0.00000	0.50000	10 (0.31)	0.00000	0.25000	0.00000
3	Farm 3	125.00%	1.00000	1.25000	0.00000	5			
4	Farm 4	107.50%	1.00000	0.70000	0.37500	5			
5	Farm 5	64.67%	1.00000	0.57851	0.06818	3 (1.09) 4 (0.06)	0.00000	0.00000	0.00000
6	Farm 6	75.00%	1.00000	0.74862	0.00138	3 (1.00)	0.00000	0.00000	0.00000
7	Farm 7	77.78%	1.00000	0.77778	0.00000	3 (1.17)	0.00000	0.00000	8.00000
8	Farm 8	85.71%	1.00000	0.42857	0.42857	4 (0.29) 10	0.00000	0.00000	0.00000
9	Farm 9	80.00%	1.00000	0.80000	0.00000	3 (0.73)	0.00000	0.00000	3.30000
10	Farm 10	128.57%	1.00000	0.00000	1.28571	4			
11	Farm 11	28.57%	1.00000	0.14286	0.14286	4 (0.19) 10	0.00000	0.00000	0.00000
12	Farm 12	13.64%	1.00000	0.11364	0.02273	3 (0.11) 4 (0.05)	0.00000	0.00000	0.00000

Fig. 4.2 Results obtained with the software EMS

An important additional benefit from this extension to the DEA model is that the set of weights is uniquely determined for the efficient units in all practical applications (Santos 1994).

We can confirm the previous results with those from the software EMS developed by Holger Scheel (2000), whose results are presented in Fig. 4.2.

We can confirm the correctness of our results, namely, that virtual inputs are always one and that the values obtained for the superefficiency are equal to those previously presented in Table 4.2.

It is worthy to note that DMUs F3, F7, and F9 place a minimal weight on output 2, while DMUs F2 and F10 choose to ignore output 1. This kind of problem is usually solved by a technique known as weight restrictions in the sense of avoiding such a flexible set of weights and incorporating value judgments.

4.3 Graphical Solution

The linear program defined by Expressions (4.1), (4.2), (4.3), (4.4), and (4.5) can be solved by the traditional graphical method if we have to deal with only 2 variables.

To reduce the problem from 3 to 2 variables, we will exploit the fact that we assume to be working under the constant returns to scale assumption, and so we will scale all data to unity input level. The new data set will be the one present in Table 4.3.

Since $x_{n1} = 1$ its multiplier $\mu_{n1} = 1$ and we have to solve the linear program just for v_{n1} and v_{n2}.

Table 4.3 Normalised data to unity input level and intersections with the axis for graphical solution

	Coefficients for the linear prog.		Intersections with the axis		
DMU	X_{n1}	$Y_{n1} = y_1/x_1$	$Y_{n2} = y_2/x_1$	v_{n1}	v_{n2}
1	1	0,500	7,000	2,000	0,143
2	1	0,200	4,500	5,000	0,222
3	1	1,000	2,000	1,000	0,500
4	1	0,800	6,000	1,250	0,167
5	1	0,636	1,500	1,572	0,667
6	1	0,750	1,500	1,333	0,667
7	1	0,778	0,667	1,285	1,499
8	1	0,600	6,000	1,667	0,167
9	1	0,800	1,000	1,250	1,000
10	1	0,500	9,000	2,000	0,111
11	1	0,200	2,000	5,000	0,500
12	1	0,125	0,500	8,000	2,000

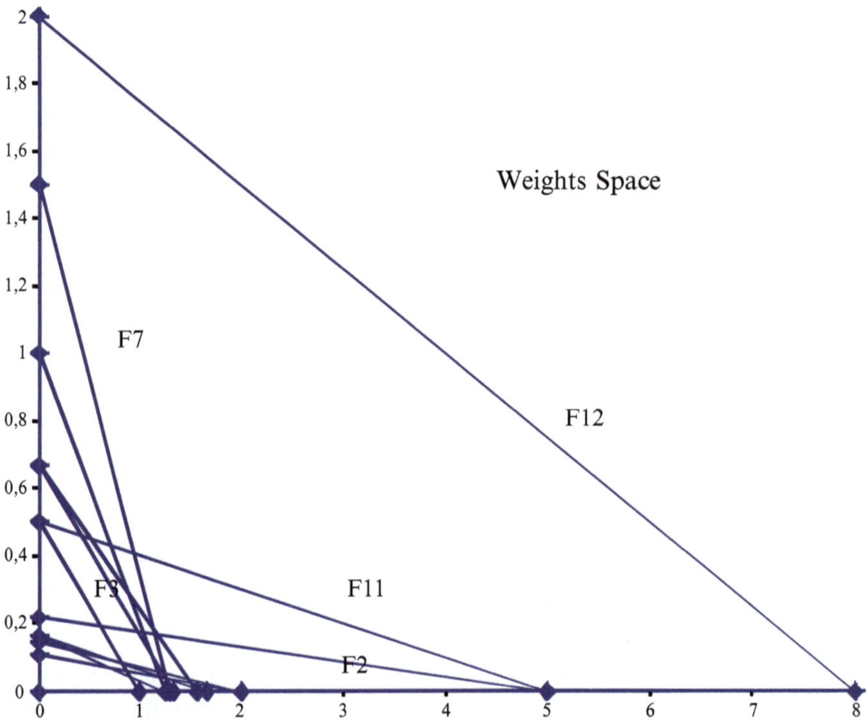

Fig. 4.3 Restrictions imposed by the 12 units in the weights space

We will just make an example with farm 12. The first constraint will be $0.125v_{n1}+0.5v_{n2} = 1$. This results in a line crossing the v_{n1} axis at $v_{n1} = 1/0.125 = 8$ and crossing the v_{n2} axis at $v_{n2} = 1/0.5 = 2$ as shown in the corresponding row of Table 4.3, and represented as the outermost line in the graphical resolution of Fig. 4.3. This is clearly a nonbinding constraint.

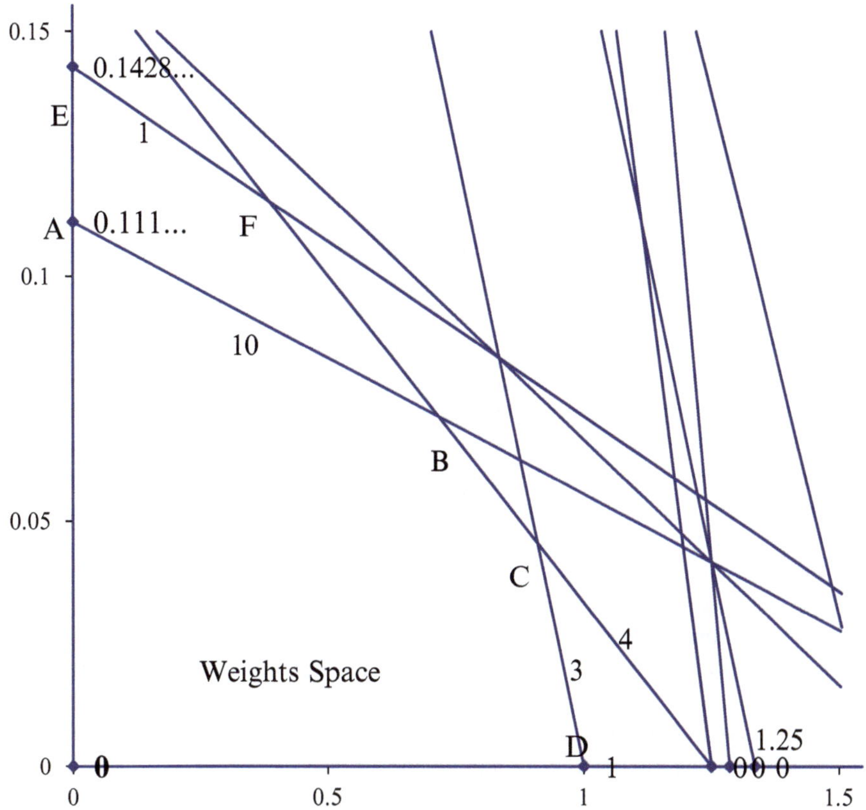

Fig. 4.4 Detail around the origin of the constraints imposed by the 12 units in the weights space

By solving graphically, we get the graph shown in Fig. 4.3 where we have 12 line segments, one for each constraint. The feasible region is the intersection of the 12 (lower) half-planes containing the origin. It is clear that inefficient units like farms 12 and 11 correspond to nonbinding constraints and that the objective function is parallel to the respective DMU constraint.

For the sake of clarity, we will expand the previous picture just to the efficient units, in order to highlight the feasible region defined by the pentagon ABCD0 (in the clockwise direction). This feasible region is always the same for inefficient units; for the efficient ones, it depends on the kind of model we are working with.

In the case of the traditional model, the feasible region is always identical, in the deleted domain case, also known as superefficiency, the constraint related to the DMU being analysed is deleted, and even a non-efficient unit can pop-up in its reference set (Santos 1994); this is the case of unit 10. In fact, its reference set is unit 1, since the feasible region on the weights space is EFCD0 (in the clockwise direction) and the objective function is parallel to the line segment 10, where it takes the unity value (Fig. 4.4).

4.4 Weight Restrictions

To avoid a given DMU "to choose" a rather unbalanced set of weights (as is the case of units F2 and F10 which ignore output 1, while F3, F7, and F9 ignore output 2), it is current practice to place some restrictions on the weights or in the virtual inputs/outputs. This is a usual way to incorporate judgement values and increase the discriminating potential of the model.

Restrictions on weights can be divided in two broad categories: relative and absolute weights restrictions. As far as we know, only linear weights restrictions have been considered in the literature; thus, we may present the weights restrictions in matrix form as $B^x \mu^T + B^y \nu^T \leq C$. In the latter expression, $\boldsymbol{B}^x \in \Re^{c \times m}$ and $\boldsymbol{B}^y \in \Re^{c \times n}$. Dimension c refers to the number of constraints.

If $C \neq 0$, we refer to them as absolute weights restrictions. The absolute weights restrictions are typically imposing a range for an individual weight. This approach was developed by Dyson and Thanassoulis in 1988 and generalised in 1991 (Roll and Cook 1991). Virtual weights restrictions introduced by Wong and Beasley in 1990 belong to this category too.

If $C = 0$, we speak about relative weights restrictions since if μ_0, ν_0 is a feasible solution, so is $k\mu_0, k\nu_0$. The class of relative weights restrictions includes, among others, the assurance region models of type I or II introduced by Thompson et al. in 1986 and 1990, respectively, as well as cone ratio DEA models from Charnes et al. in 1989.

We can say that the more general approach is to restrict the weights to belong to a closed set, being it a polytope or a polyhedral cone.

4.5 Weight Adjustment by Goal Programming

With standard DEA, it is common that many weights are null in the optimal solution. One way to avoid this situation is to place restrictions on the weights, but, in this case, weights typically used to cluster in the upper or lower limits. By including some nonlinear but convex penalty in the objective function and penalising deviations from a preference region in the weights space, it is possible to have a more uniform distribution of the weights.

This can be accomplished by the model, described by Expressions (4.6), (4.7), (4.8), (4.9), (4.10), (4.11), and (4.12) where G stands for global objective function and $P(\vec{d})$ is a penalising function of the deviations d_m and d_n of the weights from the exogenously imposed goals g_m and g_n.

The penalising function will always be a convex one, for avoiding difficulties with local minima; it is worth recalling that since the feasible region is convex and the symmetric of the objective function is also convex, a local maximum is also a global maximum (Hillier and Lieberman 1990). The definition of efficiency remains unchanged so that this model just adjusts multipliers by a penalising function appended to the objective function scaled by a constant *k*:

$$\text{Max } G_{j'} = \sum_{n=1}^{N} v_{nj'} y_{nj'} - k \times P\left(\vec{d}\right) \tag{4.6}$$

$$\text{s.t. } d_m = \mu_{mj'} - g_m \quad m = 1 \ldots M \tag{4.7}$$

$$d_n = v_{nj'} - g_n \quad n = 1 \ldots N \tag{4.8}$$

$$\sum_{m=1}^{M} \mu_{mj'} x_{mj'} = 1 \tag{4.9}$$

$$\sum_{n=1}^{N} v_{nj'} y_{nj} \leq \sum_{m=1}^{M} \mu_{mj'} x_{mj} \quad j = 1 \ldots J \tag{4.10}$$

$$\mu_{mj'} \geq \varepsilon > 0 \quad m = 1 \ldots M \tag{4.11}$$

$$v_{nj'} \geq \varepsilon > 0 \quad n = 1 \ldots N \tag{4.12}$$

We will illustrate this technique with a simple example from the previous data set with 12 DMUs.

The preferred location for the weights is around the line $0.4v_{1j'} = 4v_{2j'}$. We are not interested in justifying this choice, neither other details like the value for the constant *k*, nor the explicit kind of penalising function; we also remind that our goal is just to exemplify a way to adjust weights in a smoother way than the usual hard restriction techniques do.

Therefore, the global objective function will take the following form:

$$G_{j'} = \underbrace{\sum_{n=1}^{N} v_{nj'} y_{nj'}}_{L=\text{Efficiency}} \underbrace{- \underbrace{2 \times (0.4\, v_1 - 4\, v_2)^2}_{\text{Penality(Strictly Convex)}}}_{-k \times P(\vec{d}) \xrightarrow{\text{is}} \text{Strictly Concave}} \tag{4.13}$$

We will now illustrate the graphical solution for the evaluation of unit 10 under the constant returns to scale assumption and deleted domain technique.

In Fig. 4.5, the dotted lines correspond to constraints related to units 1, 3, and 4. The solid curves represent the isoquants of the global objective function for $G = 0.6$, 1.1, 1.6, and finally 2.1.

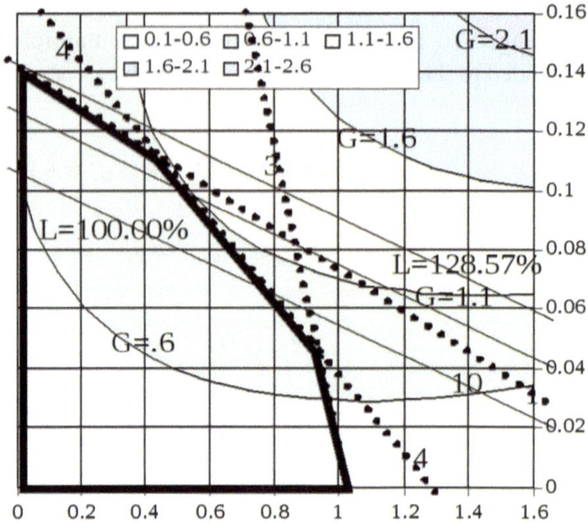

Fig. 4.5 Graphical solution for the evaluation of unit 10

We can also see 3 solid straight lines:

1. The isoquant for $L = 100.00\%$ that coincides with the constraint that was removed because of the deleted domain technique.
2. The isoquant for $L = 114.06\%$ as determined by the exact quadratic programming solution shown in Table 4.5.
3. The isoquant for $L = 128.57\%$ that equals the score value from the Superefficiency technique depicted in Fig. 4.2.

The optimal solution of the superefficiency CCR model is the basic solution defined by the intersection of the constraint relative to unit 1 and $\mu_1 \geq \varepsilon > 0$. It is interesting to note that an inefficient unit (DMU 1) is defining the optimal solution for an efficient one, a fact first published in 1994 by Santos (1994).

The solution of the quadratic programme can be obtained in a rough manner by the graphical method as exemplified in Fig. 4.5, or by the results shown in the line corresponding to DMU 10 in Table 4.5. Since the constraint relative to DMU 4 is the only binding one, its Lagrange multiplier is the only one nonzero ($\lambda_4 = 1.000$).

The optimum value for the objective function is $G = 1.070$, which occurs at the points $V_1 = 0.51$, $v_2 = 0.0982$, deviating $d = -0.188$ from the preferred linear relation between weights; as a result, we get the final value for efficiency of $L = 114.06\%$. This value is lower than the score obtained by the EMS software, since this new result has multipliers that are more desirable in the point of view of the incorporated weights preference structure, by the preferred linear relation between weights (Table 4.4).

The objective function depends on the DMU being evaluated, in the previous picture, DMU 10 was considered; now we will show the graphical solution for DMU 3.

Table 4.4 Exact quadratic programming solutions

D	ν_1	ν_2	G	d	L	λ_3	λ_4	λ_{10}
1	.71	.0714	0.857	0.000	85.71%	.000	0.24	.619
2	.68	.0732	0.465	−0.020	46.59%	.000	0.00	.464
3	.92	.0434	0.934	0.196	101.15%	.000	0.86	.000
4	.87	.0625	1.055	0.100	107.50%	.245	0.00	.790
5	.83	.0547	0.589	0.117	61.60%	.000	0.56	.000
6	.87	.0504	0.686	0.147	72.97%	.000	0.64	.000
7	.90	.0455	0.671	0.182	73.76%	.014	0.59	.000
8	.71	.0714	0.857	0.000	85.71%	.000	0.57	.286
9	.90	.0459	0.707	0.179	77.04%	.000	0.64	.000
10	.51	.0982	1.070	−0.188	114.06%	.000	1.00	.000
11	.71	.0714	0.286	0.000	28.57%	.000	0.19	.095
12	.73	.0692	0.125	0.016	12.60%	.000	0.13	.000

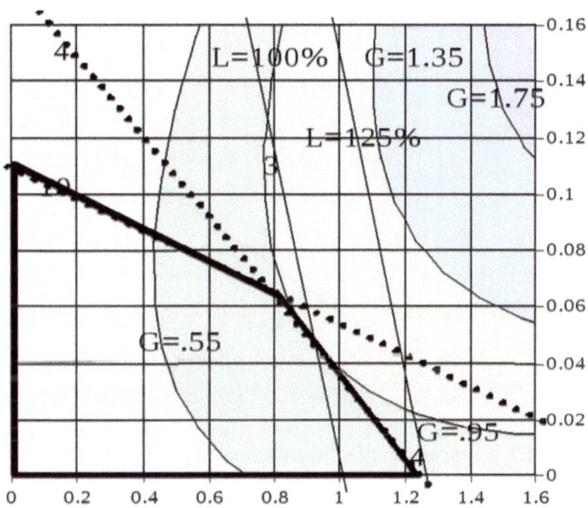

Fig. 4.6 Graphical solution for the evaluation of unit 10

In Fig. 4.6, the dotted lines correspond to constraints related to units 10 and 4. The solid curves represent the isoquants of the global objective function for $G = 0.55, 0.95, 1.35$ and finally 1.75.

We can also see 2 solid straight lines:

1. The isoquant for $L = 100.00\%$ that coincides with the constraint that was removed because of the deleted domain technique, very close to the isoquant for $L = 101.15\%$ as determined by the exact quadratic programming solution shown in Table 4.5.
2. The isoquant for $L = 125.00\%$ that equals the score value from the Superefficiency technique depicted in Fig. 4.2.

Table 4.5 Statistical descriptors of the results for the EMS software (CCR model, minimization of inputs)

	Eff.	μ_1	μ_2	μ_3	μ_4	V_1	V_2
Average	84%	.03	.01	.01	.02	.09	.18
Standard dev.	19%	.03	.02	.02	.03	.08	.09
Coef. of dar.	.222	.99	1.53	1.98	1.31	.87	.51
Maximum	127%	.10	.10	.10	.10	.28	.32
Percentile 75%	97%	.06	.02	.01	.04	.14	.24
Percentile 50%	85%	.03	.00	.00	.01	.06	.20
Percentile 25%	70%	.00	.00	.00	.00	.03	.15
Percentile 4%	55%	.00	.00	.00	.00	.00	.00
Minimum	44%	.00	.00	.00	.00	.00	.00

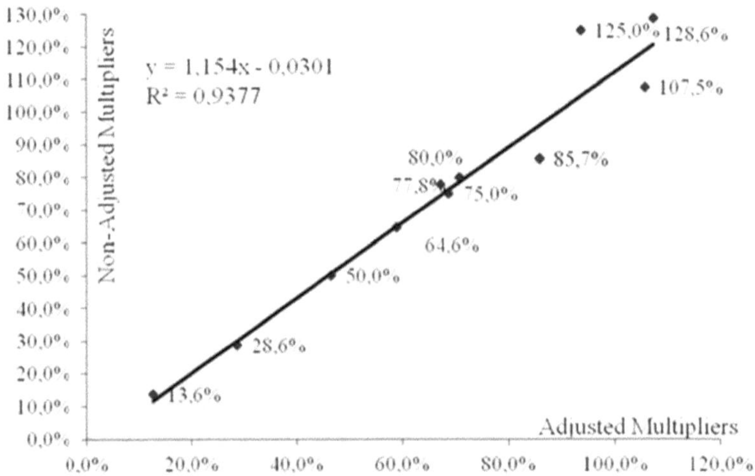

Fig. 4.7 Comparison of efficiency scores

It is interesting to remark that the quadratic objective function isoquants have a different orientation from those depicted in Fig. 4.5, since the slope of the linear traditional linear objective functions were different too.

Since the constraint relative to DMU 4 is the only binding one, its Lagrange multiplier is the only one nonzero ($\lambda_4 = 0.857$).

The optimum value for the objective function is $G = 0.934$, which occurs at the point $v_1 = 0.92$, $v_2 = 0.0434$, deviating $d = 0.196$ from the preferred linear relation between weights; as a result, we get the final value for efficiency of $L = 101.15\%$. Again, this value is lower than the score obtained by the EMS software, since this new result has multipliers that are more evenly distributed; around our goal, the line $v_2 = 0.1v_1$. The results without weight adjustment were an efficiency score of $L = 125.00\%$ but with a weight pattern neglecting output 2 ($v_1 = 1.00$, $v_2 = 0$).

We can compare the score efficiencies obtained by the two models. The results are as depicted in Fig. 4.7.

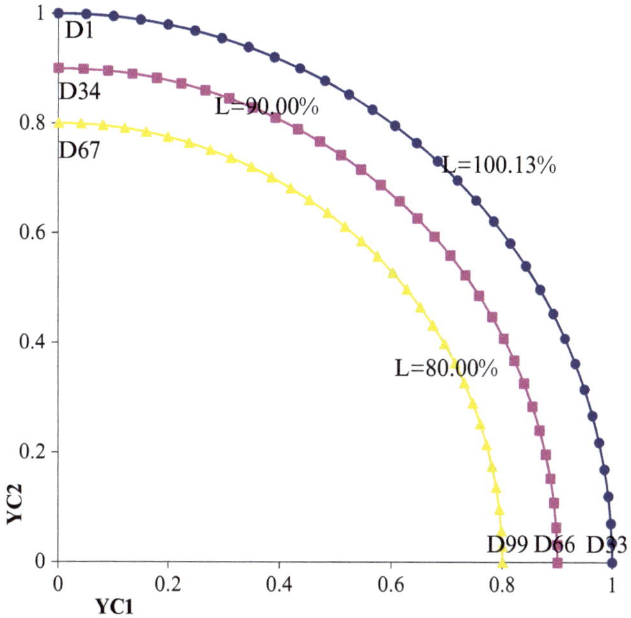

Fig. 4.8 Data for an example of absolute weights adjustment

The dotted line is the line defined by $y = x$ (restricted to the first orthant), to make it easy to see that all points are above or on that line, meaning that efficiency scores obtained with weight adjustment are not bigger than the ones obtained without adjustment. There are 3 DMUs (1, 8, and 11) which efficiency score remained unchanged, as a result of the fact that their optimal weights as shown in the EMS results in Fig. 4.2 were already complying with our later specification of staying on the line $v_2 = 0.1v_1$.We can get the full detailed results for the other remaining 10 DMUs by Table 4.5 and locating their optimal weights in Figs. 4.5 or 4.6.

It is now worth noting that DMUs 1 and 8 have exactly the same score and the same optimal weights, since the relation $v_2 = 0.1v_1$ implies an intrinsic marginal rate of substitution of y_1 by y_2 of $v_{1/}v_2 = 10$, which is exactly the case for their outputs. The fact that two of its Lagrange multipliers are the only nonzero ones (λ_4 and λ_{10}) means that its optimal weights are in the intersection of the two straight lines relative to the binding restrictions of DMUs 4 and 10.

Now we will introduce a last graphical example of the solution of the CCR model of the data set presented in Fig. 4.8 where we consider one input $X = 1$ and two outputs YC_1 and YC_2 whose name comes from the appearance of the 99 points in circular layers in the first orthant. In fact, the solver for nonlinear programs we are using is limited to 100 constraints.

In Fig. 4.9, we also present the results of the EMS software under deleted domain and constant returns to scale assumption.

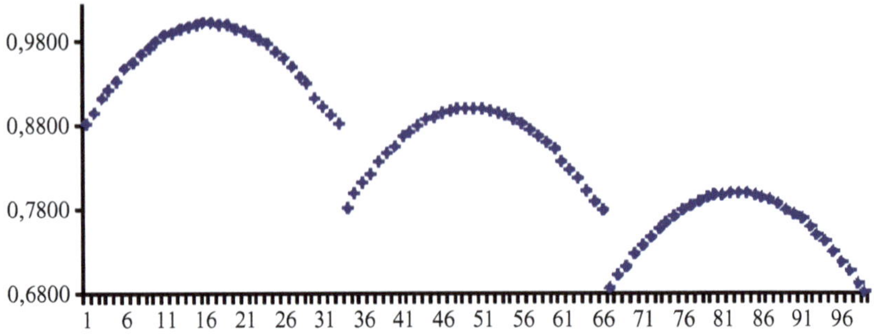

Fig. 4.9 Efficiency scores for absolute adjustment

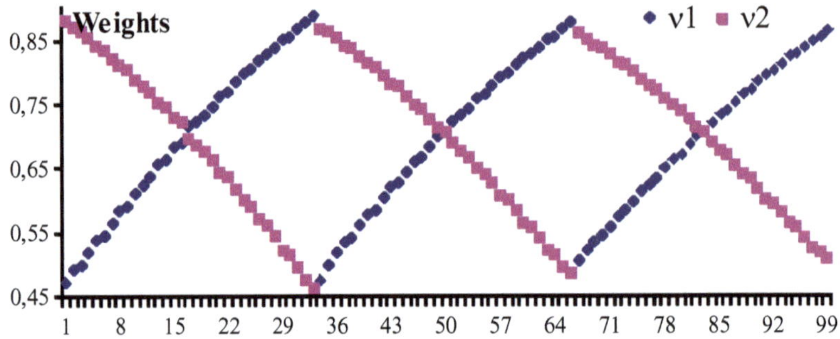

Fig. 4.10 Optimal weights distributing around 0.6

This time the deviation at the penalising function was defined as the squared distance from the average of the standard DEA weights normalised by its standard deviation:

$$P\left(\vec{d}\right) = \left(\frac{(v_1 - \mu_{v_1})}{\sigma_{v_1}}\right)^2 + \left(\frac{(v_2 - \mu_{v_2})}{\sigma_{v_1}}\right)^2 \tag{4.14}$$

The meanings of the parameters in Expression (4.14) are as follows: v_j is the unknown weight for YC_j, μv_j is the average of the weight v_j determined by solving the standard DEA CCR model under deleted domain, and σv_j is the standard deviation of the weight v_j found by solving the standard DEA CCR model under deleted domain.

The efficiency scores for this absolute adjustment are shown in Fig. 4.9, where the highest efficiency score of 100.13% is attained by a single DMU. In fact, the 17th DMU has the same weights than those obtained by the standard minimization of inputs DEA CCR model under deleted domain.

In Fig. 4.10, we present the resulting weights that instead of clustering at the upper and lower limits of its absolute weights restrictions, as what happens with traditional weights restriction techniques, now they spread around the central previously defined value $v = (06, 0.6)$.

We do not know of any other public work that can accomplish such versatility as this technique.

This method can be further enhanced by the introduction of special types of non-convex penalty functions like Chebyshev polynomials or maximally flat polynomials.

4.6 Application to a Case with Artificially Generated Data

Finally, we will apply this multiplier adjustment technique to a case with two outputs and four inputs, with simulated data.

4.6.1 Data-Generating Process

When generating the data for the experiments reported in this section, we considered two factors: production technology and inefficiency distribution.

Production Technology: Consider a two outputs and four inputs production technology. This was obtained through the generation of a single aggregate output incorporating an inefficiency stochastic factor and later split in two outputs as suggested in 1997 by Durchholz and Barr.

The single Cobb-Douglas aggregate output and four inputs production technology is specified in terms of its efficient production function $z = f(x_1, x_2, x_3, x_4)$, where z represents the maximum aggregate output that can be produced from the levels x_1, x_2, x_3, and x_4 of the four inputs. Specifically, the following shifted Cobb-Douglas production function was used:

$$z = a_0 \prod_{i=1}^{4} (x_i - \alpha_i)^{\beta_i} \tag{4.15}$$

where α_0 is a constant scale factor, α_i is the shift from the origin for input i, β_i is the factor elasticity for input i, and z is the single Cobb-Douglas aggregate output.

If $\sum \beta_i = 1$, then only constant returns to scale exist in the production process. For $\sum \beta_i < 1$, decreasing returns to scale are present, while $\sum \beta_i > 1 \sum$ indicates increasing returns to scale.

In DEA, increasing returns to scale are not used because the function results in only a few DEA efficient points, since for higher levels of input than those defined by the most productive scale size (MPSS), the production possibilities set (PPS) is the polyhedral cone from the CCR model. Although this does not pose a problem, it does not represent realistic data.

We made $\alpha_0 = 1$, $\alpha_1 = \alpha_2 = \alpha_3 = \alpha_4 = 5$, the input levels x_1, x_2, x_3, and x_4 were generated randomly from four independent uniform probability distributions over the interval [10, 20], and the coefficients β_2, β_3, and β_4 were randomly generated from independent uniform probability distributions over the interval [0.20, 0.25]. Since the sum of β_1, β_2, β_3, and β_4 is less than one, the production function in Expression (4.15) satisfies the DEA-BCC models assumption of a strictly concave production function, while the shifts $\alpha_1 = \alpha_2 = \alpha_3 = \alpha_4 = 5 > 0$ allow both increasing and decreasing returns to scale to prevail.

Inefficiency distribution: The logarithm of the inefficiency, $u = ln\,\theta$ from a half-normal distribution $|N(0;\sigma_u^2|$, where the parameter σ_u^2 itself is drawn from a uniform distribution on the interval [0, 0.1989], was generated. The range of values for the distribution of σ_u^2 is chosen in such a way that the mean efficiency given by $E(h = 1/\theta) = exp\,(-\sigma_u^{\sqrt{2/\pi}})$ is between 0.7 and 1.0.

Simulated Observations: First random values for β_i between 0.2 and 0.25 and a value σ_u^2 between 0 and 0.1989 were generated. Next, we simulated 90 observations of the four inputs $\boldsymbol{x_1}$ through $\boldsymbol{x_4}$ between 10 and 20. Those values were then substituted into the efficient production function specified in Expression 21 to obtain the corresponding values $z_j = f(x_{ij})$ for the efficient output quantity. Then, we randomly generated the logarithm of actual inefficiency values $u_k = ln\theta_k$, for each one of the 90 observations from the half-normal distribution $|N(0;\sigma_u^2|$.

Finally, we obtained the values for observed aggregated output quantities y_j as $y_j = f(x_{ij})/\exp(u_j)$ and the true efficiency value $h_k = 1/\exp(u_k)$.

Once the single aggregate output level y was calculated, the two individual output levels were determined by assigning each individual output as a percentage of the aggregate. The percentages for each individual output were drawn from normal distributions with predetermined means and standard deviations. The means of the normal distributions were chosen so that the percentages sum up to one. In our speçific case, we chose each normal distribution to have a mean of 0.5 and a standard deviation of 0.1.

The results from the unbounded model solved by the EMS software are summarised in Table 4.5.

From the comparison between Tables 4.5 and 4.6, we notice that the statistical descriptors of the efficiency scores in our model are always lower. This was expected since penalties may occur and our global objective function never exceeds the traditional linear one.

The average efficiency score in the classical CCR model is 84%, as was expected from the assumptions about the half-normal distribution of the inefficiency.

The standard deviation of any multiplier is lower than the corresponding one in the nonadjusted case (this could originate from the decrease in the average values that lead us also to compute the coefficient of variation, confirming that, even in relative terms, the weights are not as spread as in the original model).

The same conclusion also holds to any other measure (maximum and percentiles).

Only when it comes to the minima of μ_4 and v_2, we have a tie, but even in this case, it is sufficient to take into account the values for the lower percentiles. In fact, μ_4 vanishes 35 times in the traditional linear model, in opposition to just 5 zeros in our model.

Table 4.6 Statistical descriptors of the results for our new model (CCR model, minimization of inputs)

	Eff.	μ_1	μ_2	μ_3	μ_4	ν_1	ν_2
Average	74%	.035	.012	.007	.018	.074	.176
Standard dev.	16%	.014	.005	.003	.011	.051	.064
Coef. of var.	.221	.403	.421	.393	.633	.692	.367
Maximum	123%	.066	.024	.014	.042	.247	.276
Percentile 75%	90%	.045	.015	.008	.027	.098	.218
Percentile 50%	81%	.034	.011	.007	.017	.052	.192
Percentile 25%	65%	.023	.009	.005	.008	.042	.155
Percentile 4%	48%	.015	.003	.002	.000	.027	.035
Minimum	42%	.012	.002	.001	.000	.002	.000

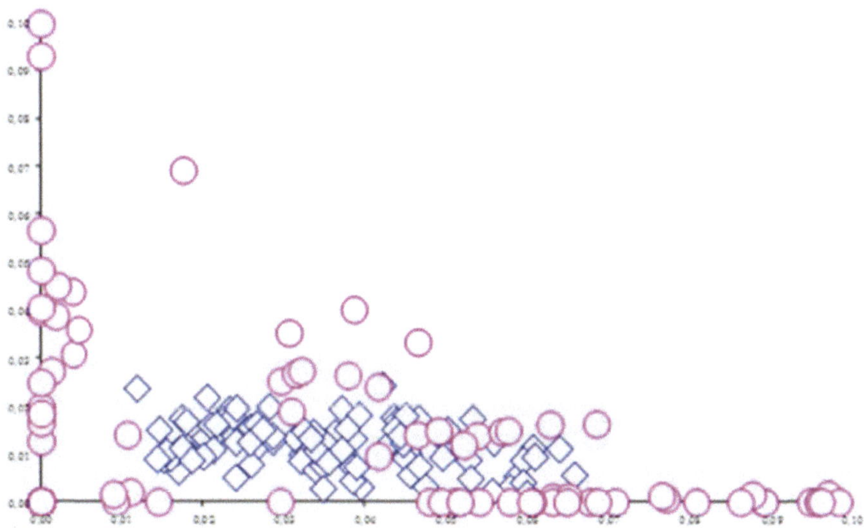

Fig. 4.11 Distribution of μ_1 (in the vertical axis) and μ_2 (in the horizontal axis)

In the case of ν_2, the proportion is similar: There are only two null weights in the adjusted case against 13 on the classic one.

It should be noted that it would be easy to avoid the occurrence of null weights simply by increasing the steepness of the convex penalty function, although this was not our choice, since it could lead to a point that no DMU at all would reach the 100% efficiency score

In an effort to show our results in the 6-dimensional weights space, we illustrate in Figs. 4.11, 4.12, and 4.13 its projections on the bidimensional spaces. In these figures, the circles represent the weights for the traditional model and the diamonds correspond to those obtained by our model.

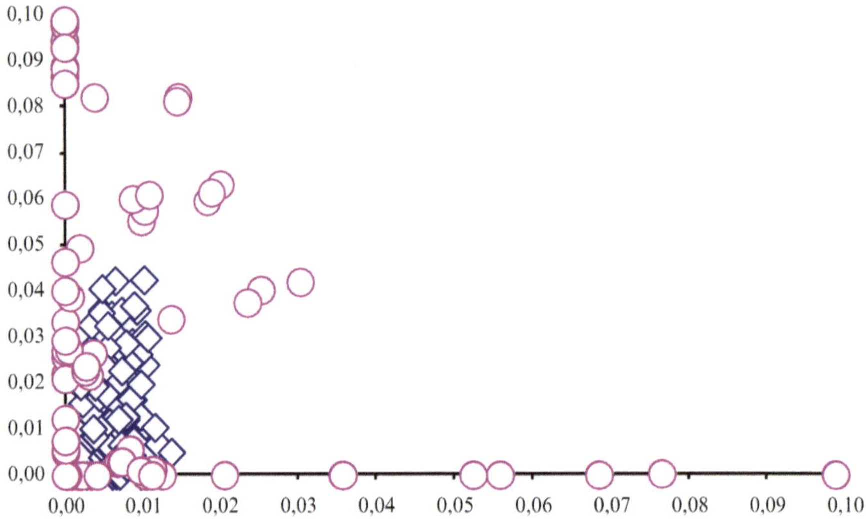

Fig. 4.12 Distribution of μ_3 (in the vertical axis) and μ_4 (in the horizontal axis)

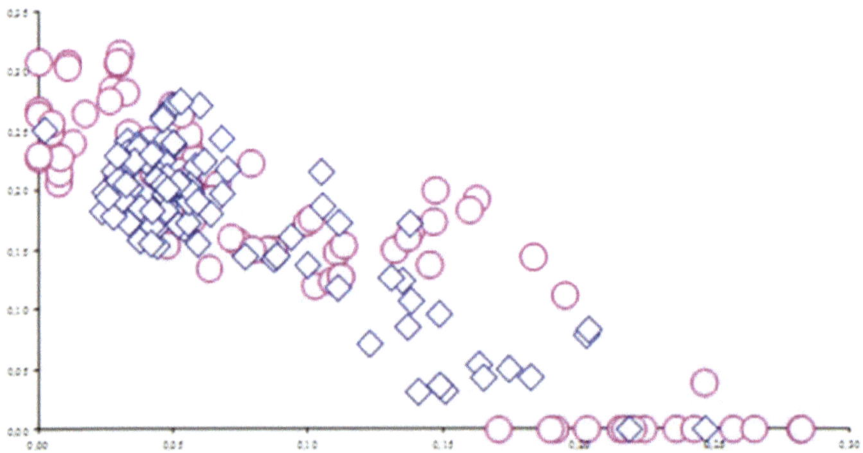

Fig. 4.13 Distribution of ν_1 (in the vertical axis) and ν_2 (in the horizontal axis)

In Fig. 4.11, we can confirm that μ_1 and μ_2 cluster over the both axes, if we do not apply our adjustments on multipliers. In fact, μ_1 vanishes 31 times and about μ_2 this happens 43 times.

The considerations made about Fig. 4.11 also apply to Fig. 4.12, but here μ_3 has 38 zeros and μ_4 has 35. As a result of applying our technique, only μ_4 kept some zeros, but nevertheless its number dropped to 5. It is worth mentioning that not only did we avoid the occurrence of zeros but we also reduced its maximal values.

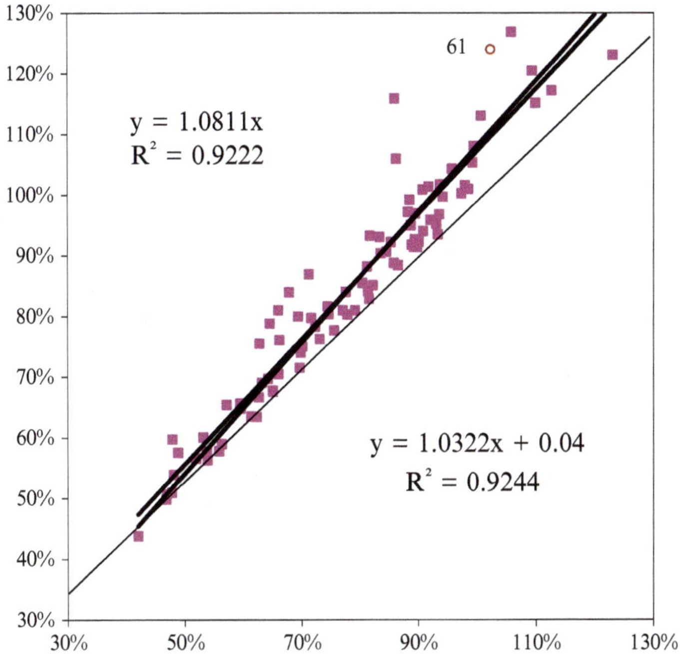

Fig. 4.14 Representation of the score obtained by the EMS software and our efficiency values

In Fig. 4.13, since we are dealing with only two outputs, the benefits of our method are not as evident as in the input case. Regardless of that, from the initial 13 DMUs that presented a null weight in output Y_2, this number decreased to only 2 of the initial 13, namely, DMU 41 and DMU 61.

If we had had more outputs, our technique would have been more useful, in the sense that the number of zeros to reduce would have been greater.

We tried to make this representation in several graphic ways, but this one seemed to us to be the best. We also investigated if the use of a data reduction technique like factor analysis could be of some help, but the correlation matrix did not allow for an easy alternate representation of the data.

In Fig. 4.14, we plot the results for both the score obtained by the EMS software and our efficiency values. We notice that our values never exceed those from the CCR model and that, in some cases, there is a substantial reduction on the original value as is the case by instance of DMU 61 whose efficiency value dropped from an original value of 124–102%.

Despite this fact, it is important to remark the existence of a strong correlation between the two variables, and so we made a linear regression on it.

Although the model with a constant leads to a higher determination coefficient of 0.9244, this constant has a p value of 0.11, and therefore, it is not significant. Thus, we conclude that the EMS scores exceed in 8% those from our model.

4.7 Conclusions

This chapter introduces superefficiency and a new way of adjusting weights, a matter that has already deserved many publications in the data envelopment analysis field. This new method adds greater flexibility to the weight restrictions techniques. It is not our concern to present the potentialities or the details of weight restrictions. This matter is already extensively covered in the existing literature, namely, on how to set up the specific values for the restrictions.

Since we are dealing with a nonlinear concave objective function, we have the possibility to locate the optimal set of weights in a continuous way in contrast to the linear case where optimality always occurs at a vertex of the feasible solution set.

We used the simplest convex penalty function for the sake of clarity, but other convex functions like maximally flat polynomials or Chebyshev polynomials could also be used.

Even the convexity restriction can be dropped leading to more complicated programs that can be used for instance in discriminating analysis.

References

Andersen P, Petersen NC (1993) A procedure for ranking efficient units in data envelopment analysis. Manag Sci 39(10):1261–1264

Banker RD, Charnes A, Cooper WW (1984) Some models for estimating technical and scale inefficiencies in data envelopment analysis. Manag Sci 30(9):1078–1092

Boussofiane A, Dyson RG, Thanassoulis E (1991) Applied data envelopment analysis. Eur J Oper Res 52:1–15

Charnes A, Cooper WW, Rhodes E (1978) Measuring the efficiency of decision making units. Eur J Oper Res 2(6):429–444

Charnes A, Cooper W, Wei Q, Huang Z (1989) Cone ratio data envelopment analysis and multi-objective programming. Int J Syst Sci 20:1099–1118

Durchholz M, Barr R (1997) Parallel and hierarchical decomposition approaches for solving large-scale data envelopment analysis models. Ann Oper Res 73:339–372. doi:10.1023/A:1018941531019

Dyson RG, Thanassoulis E (1988) Reducing weight flexibility in data envelopment analysis. J Oper Res Soc 39(6):563–576

Hillier FS, Lieberman GJ (1990) Introduction to operations research. Mc Graw Hill, New York

Roll Y, Cook W (1991) Controlling factor weights in data envelopment analysis. IIE Trans 23(1):2–9

Santos J (1994) Ordenação de unidades eficientes por técnicas de data envelopment analysis. Unpublished MsC. thesis, Instituto Superior Técnico, Technical University of Lisbon

Santos J (2008) Issues of universal feasibility and multiplier adjustment in data envelopment analysis (DEA) with an application. Unpublished PhD thesis, Universidade de Évora

Scheel H (2000) EMS: efficiency measurement system user's manual version 1.3. http://www.microtheory.uni-jena.de/download/ems.pdf

Thompson RG, Singleton FD, Thrall RM, Smith BA (1986) Comparative site evaluations for locating a high-energy physics lab in Texas. Interfaces 16(6):35–49

Thompson RG, Langemeier LN, Lee CT, Thrall RM (1990) The role of multiplier bounds in efficiency analysis with application to Kansas farming. J Econ 46:93–108

Wong Y-HB, Beasley JE (1990) Restricting weight flexibility in data envelopment analysis. J Oper Res Soc 41(9):829–835

Part II
Farm Efficiency Applications

Chapter 5
An Application of Data Envelopment Analysis (DEA) in Azores Dairy Farms

Emiliana Silva, Amílcar Arzubi, and Julio Berbel

Abstract This research measures the Azores dairy farms' technical efficiency by applying a non-parametric efficiency analysis to a panel data of 122 dairy farms from the Azores, Portugal, for 1996. The analysis used DEA with constant and variable returns to scale models, with an input-oriented model approach. Two outputs (milk production and subsidies) and three inputs (agricultural area, number of dairy cows and variable and fixed cost) were considered relevant. The results suggest that the average technical efficiency is very low (66.4%) compared with published research data, and only a few (7%) dairy farms were found to be efficient.

Keywords Azores • Dairy efficiency • Data envelopment analysis • Models • Non parametric

5.1 Introduction

Azores is Portuguese insular territory with a population about 250,000 inhabitants, where main economic activity is dairy farming. Dairy policy depends on CAP (Common Agricultural Policy) of the European Union. Different authors have done analysis of dairy farming by using different methodologies; this chapter tries to

E. Silva (✉)
CEEApIA, Departamento de Ciências Agrárias, Universidade dos Açores, Rua Capitão João de Ávila, 9700-042 Angra do Heroísmo, Açores, Portugal
e-mail: emiliana@uac.pt

A. Arzubi
Lomas de Zamora University, Lomas de Zamora, Argentina

J. Berbel
ETSIAM, Department of Agricultural Economics, Córdoba University, Córdoba, Spain

A.B. Mendes et al. (eds.), *Efficiency Measures in the Agricultural Sector: With Applications*, DOI 10.1007/978-94-007-5739-4_5,

study the sector from the agricultural economics' point of view based upon firms' typology behaviour.

The methodology in this chapter is based in the estimation of technical efficiency of dairy farm, using random selection of 122 dairy farms from the database (1996) of the European Union, Farm Accountancy Data Network (FADN).

Data envelopment analysis (DEA) is used to estimate the efficiency. Charnes et al. (1978) developed the method from the earlier works by Farrell (1957). This non-parametric method has been used to estimate the efficiency in the organisative units in several areas; Cooper (1999) presents a general review of the past development and future possibilities of the methodology.

DEA involves the use of technical linear programming to construct a non-parametric piecewise surface (or frontier) over data and also to enable the calculation of an efficient firm relative to its surface (Coelli 1996). Any farm that lies below the frontier is considered to be inefficient. DEA permits to construct a best-practice benchmark from the data on inputs and outputs (Jaforullah and Whiteman 1999).

We will use the linear programming (LP) approach, but the reader should remember that another alternative approach is the use of statistical parametric techniques, based upon econometric methods, to construct a stochastic frontier.

DEA involves the concept of efficiency, and Farrell (1957) divided the efficiency into (1) technical efficiency and (2) allocative efficiency. The technical efficiency measures the maximum equiproportion reduction in all inputs and still allows continued production of given outputs. The technical inefficiency measures the magnitude. The allocative efficiency reflects the ability of a firm to use the inputs in optimal proportions, given their respective price. These two concepts form the concept of economics efficiency (Coelli 1995). The allocative inefficiency measures the magnitude of consequent loss. Similar considerations are applied to economics efficiency and inefficiency.

Therefore, the overall measure of technical efficiency can be desegregated into three components: (1) pure technical efficiency due to producing within an isoquant frontier, (2) congestion due to overutilisation of inputs and (3) scale efficiency due to deviations from constant returns to scale (Weersink et al. 1990).

Previous applications of DEA dairy farming were made in order to estimate the efficiency by different researchers in different parts of the world (Table 5.1). We have summarised some recent papers on the subject.

Recently, Arzubi and Berbel (2002) estimate the technical dairy efficiency with DEA using 35 farms in Argentina. The average technical efficiency was 78.2%, and about 11.4% of all firms were efficient to constant returns to scale. They divide the global technical efficiency (CRS) into two components, and they observed that there is a bigger pure technical inefficiency (16.5%) compared with the scale inefficiency (6.1%). They explained this fact, with the 22 firms that were operating to increasing constant returns to scale and 8 were operating decreasing returns to scale.

Also, Reinhard et al. (2000) estimated the efficiency measures for 163 highly specialised Dutch dairy farms from FADN. They compared two methods for calculating the efficiency, namely, the stochastic frontier analysis and DEA. They

Table 5.1 Dairy farms DEA efficiency researches

	Country	Year	Technical efficiency (%)	Farms number	Efficient farms (%)
Arzubi and Berbel (2002)	Argentina	1997/1998	78.2	35	11.4
Jaforullah and Whiteman (1999)	New Zealand	1996	83.0	264	19
Fraser and Cordina (1999)	Australia	1995	90.5	50	24
		1996	90.8	50	20
Cloutier and Rowley (1993)	Canada	1988	88.0	187	1
		1989	91.0	187	21
Gonzalez Fidalgo et al. (1996)	Spain	1991	78.0	133	5

concluded that both methods can estimate environmental efficiency scores, but they differ according to the methods used.

In New Zealand, Jaforullah and Whiteman (1999) examined the relationship between size and efficiency of dairy farms. The average technical efficiency was estimated at 89%, and the results in general support a policy of encouraging increasing farm size. In this study, 19% of 264 dairy farms were found to be efficient.

DEA was used by Fraser and Cordina (1999) to assess technical efficiency of 50 dairy farms in Northern Victoria, Australia, over the periods 1994/1995 and 1995/1996. The average technical efficiency was 90.5 and 90.8% for 1994/1995 and 1995/1996, respectively. And 24% of the dairy farms were efficient for the period 1994/1995 but decreased 20% in the subsequent period.

Cloutier and Rowley (1993) applied DEA to measure technical efficiency of 187 dairy farms in Quebec, Canada, for the years 1988 and 1989. Their DEA model was based on the assumption of constant returns to scale. The technical efficiency of the Quebec dairy farms was 88 and 91% for 1988 and 1989, respectively, and 15 and 21% of dairy farms were efficient. They suggest that their results show that larger farms are much more likely to appear efficient than smaller ones.

In Spain, Gonzalez Fidalgo et al. (1996) estimated the technical dairy efficiency of a panel with 133 dairy farms in Asturias (northern Spain). The average global technical efficiency was around 78%, but only a few (5%) were efficient.

Färe and Whittaker (1995) used the non-parametric methods to estimate the efficiency of 137 dairy farms. They also took into account the decomposition production into subproduction processes with the outcome that a larger number of variables may be useful to be included in the model. They compared the efficiency with the dimension of the dairy farms (small, medium and large). The results show that using intermediate products increases the efficiency in any dimension of dairy farm considered.

In Canada, Weersink et al. (1990) estimated the technical efficiency of 105 dairy farms in Ontario. As a result, they found 43% of the dairy farms to be efficient, and their average technical efficiency laid around 91.8%.

As we can see, DEA has been widely used in dairy farming analysis. We can find many other applications to agricultural economics such as Reinhard and Thijssen (2000) who estimated the nitrogen efficiency in 434 dairy farms in Germany. They used a shadow price, and the mean nitrogen efficiency was around 56%, and the mean input-oriented technical efficiency was 84%.

On the other hand, some researches, Tauer (2001), Maieta (2000), Brummer and Loy (2000), Reinhard et al. (1999), Alvarez and Gonzalez (1999) and Hallam and Machado (1996), use parametric methods (econometric) to estimate dairy farms efficiency. Also Tauer (1988) used DEA to evaluate the efficiency of some New York dairy farms using the Malmquist productivity index.

In this study, DEA was selected because it offers the opportunity of including more than one output and permits the relationship between all inputs and outputs simultaneously. DEA also yields a relative measure of efficiency more than the frequently reported partial indicators of farm efficiency such as milk production per hectare or milk production per cow. Finally, DEA does not require a parametric specification of a functional form to define the frontier.

The major limitations of DEA are that it is conceptually difficult to separate the effects of uncontrollable environmental variables and error measurements from the effect of differences in farm management and the presence of outliers. By using the stochastic frontier methodology, one can find the causal factors that may solve this.

5.2 Material and Method

The Azorean data of FADN-1996 permits to observe that, in general, dairy farms are small, and most of them belong to the owners. The average agricultural area is around 23.7 ha (more than 85% is pasture), and the average number of dairy cows is about 23 per farm. System of production is primarily based on grazing, and the main product is milk (84%). Most expenses go on concentrates (27%), annual depreciation (13.6%), rents (10.6%) and fertilisers (9.8%) (Silva 2001).

In the former years, the European Union had national programmes to increase the farms' areas and dairy production. In the period between 1986 and 1996, the milk production increased around 70.9% as a result of the "milk quota systems" of the European market regulation.

The database was characterised by three types of grazing systems defined by Silva (2001). The first group of dairy systems was extensive (cows per hectare smaller than 1.4), the second group of dairy farms was medium-intensive system (1.4 a 2.4 cows per hectare) and the third was intensive (bigger than 2.4 cows per hectare).

The dairy farms' technical efficiency was analysed by Data Envelopment Analysis Computer Program (DEAP) developed by Coelli (1996). This program is based on the optimisation model (5.1) developed by Charnes et al. (1978); it is considering the product of input components $V_i X_{ij}$ as a constant (K):

$$\text{Max } E_j = \sum_{1}^{n} U_i Y_{ij}$$

$$\text{s.t.}:$$

$$\sum_{1}^{m} V_i X_{ij} = K$$

$$\sum_{1}^{n} U_i Y_{ij} - \sum_{1}^{m} V_i X_{ij} \leq 0$$

$$U_i, V_i \geq 0 \tag{5.1}$$

Y_{ij} is the level of output i used by decision-making unit j, X_{ij} is the level of input i used by decision-making unit j and U_i and V_i are the non-negative variable weights associated with the solution of decision-making unit j, of outputs and inputs, respectively.

In our model, two outputs were considered: y_1, litres of dairy produced, and y_2, subsidies received by farm. On the other hand, three groups of inputs were included: x_1, agricultural area (hectare), x_2, dairy herd (number) and x_3, dairy variables costs, i.e. fertiliser, feeding (concentrates, pasture and others) and fixed costs, labour, annual depreciation of buildings and machinery and paid rents (euros).

By adjusting this model to our case, we measure efficiency, E_j, like this:

$$E_j = \frac{U_1 \text{ litres of milk} + U_2 \text{ euros of subsidies}}{V_1 \text{ agriculture area} + V_2 \text{ cows number} + V_3 \text{ costs}} \tag{5.2}$$

Equation (5.2) $E_j = 1$ means the dairy farms are efficient when compared with all the other firms, and when it is smaller than one, the dairy farms are inefficient.

The constant returns to scale model (CRS) corresponds to the original model developed by Charnes et al. (1978) that assumes all firms were operating at an optimal scale. Later Banker et al. (1984) suggested a model extending the original, in which the variable returns to scale (VRS), change the linear programming by incorporating convexity limitations (restrictions). This change permitted the division of technical efficiency (or global technical efficiency) into pure technical efficiency and scale efficiency.

If the CRS and VRS are run operating with some data and if there is some difference between firms, that is due to scale inefficiency. That inefficiency must be calculated from the difference between CRS and VRS. The pure technical efficiency coincides with VRS. The scale inefficiency, can be explained as a result of the fact that scale level is not optimal (when CRS = 1). The global technical efficiency (in CRS) is the product of pure technical efficiency and scale technical efficiency. When not all decision-making units are operating at the optimal scale, it will result in technical efficiency, which can be confused with scale efficiency. The use of the VRS specification will permit the calculation of technical efficiency devoid of these scale efficiency effects.

Many studies have separated the technical efficiency scores obtained from a CRS DEA into two components: (1) one due to scale inefficiency and (2) another one due to "pure" technical inefficiency (Coelli 1996). If there is a difference in the two technical scores for a particular decision-making unit, then this indicates that the decision-making unit has scale inefficiency and that scale inefficiency can be calculated from the difference between VRS and CRS technical efficiency scores.

The FADN database includes also mixed farming; therefore, before applying the DEA method, it was necessary to extract pure dairy farms from the global database. The selection of these farms was done as follows:

1. Select the farms with OTE (Technical and Economics Orientation) 4.1, 4.2 and 4.3. This classification, provided by the European Union for all agricultural farms, has to do with ruminants.
2. Refine the FADN database and remove errors in the data.
3. Exclude those farms whose dairy production was below 3,500 l per cow each year.
4. Eliminate the dairy farms whose specialisation ratio was more than 0.66. Avillez (1991) defined this ratio for the Azorean farms. It is defined by the following formula (5.3):

$$\text{Specialisation} = \frac{\text{Total cows (meat + dairy)} - \text{dairy cows}}{\text{dairy cows}} \tag{5.3}$$

Finally, we apply DEA to 122 pure dairy farms in the Azores.

5.3 Results and Discussion

The main results are (Table 5.2):

- Only nine dairy farms were efficient, thus representing a 7.4% of the total number of farms.
- The average technical efficiency was 0.664. It is possible to produce the same amount of milk while saving approximately 33.6% of resources (or inputs).

These results showed that the Azorean dairy efficiency could be improved.

Table 5.2 Statistics of the Azorean dairy farms (technical efficiency: increasing IRS and decreasing DRS, and constant)

	CRSTE	VRSTE	SCAL
Mean	0.664	0.782	0.855
Standard deviation	0.154	0.160	0.123
Maximum	1	1	1
Minimum	0.286	0.375	0.491
Efficient farms	9	25	9
IRS dairy farms	–	–	88
DRS dairy farms	–	–	25

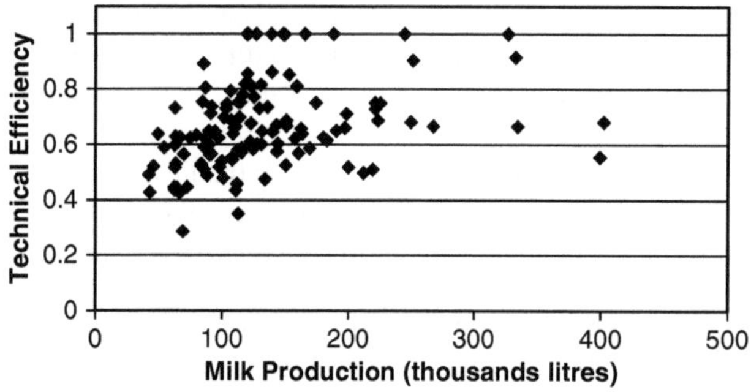

Fig. 5.1 Milk production and technical efficiency in Azores dairy farms

Fig. 5.2 Milk costs per litre and milk production per hectare

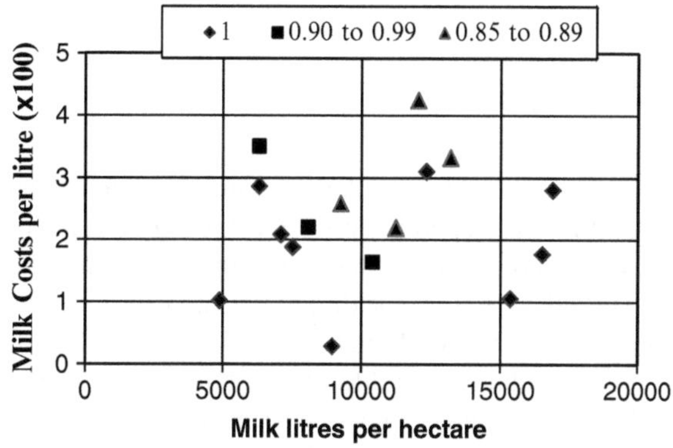

Technical efficiency from variable returns to scale model increased to 0.782, and scale efficiency is 0.855.

Scale inefficiency (14.5%) may occur due to an operation below the optimal scale, as a result of the fact that a 72.1% of dairy farms operate at increased returns to scale. And a 20.5% were operating at decreasing returns to scale.

In Fig. 5.1, it can be observed that the efficient farms exist in any level of milk production. There is no direct relation between the technical efficiency and the milk production. However, most of the efficient farms are grouped within the 100 and 200 thousands litres of milk stratum.

Figure 5.2 shows that efficient farms have reduced costs of milk production per hectare. Another factor to be considered is that efficiency does not seem to be related to the amount of milk produced per hectare. This means that there are efficient dairy farms at any level of milk production per hectare. Larger dairy farms would be expected to have higher efficiency levels as mentioned by Tauer (2001).

Comparing efficiency in-groups of Azorean dairy farms (intensive, extensive and intermediate) as defined by Silva (2001), we found that there is not one production system with a higher number of efficient farms than the other. This means that the efficiency does not depend on the intensity of grazing but probably on the importance of some fixed costs (equipment depreciation) and some variable costs (animal feeding). Silva (2001) found the animal feeding and equipment depreciation in the Azorean dairy farms to be of great importance, about 27 and 13.6% of total cost, respectively.

In the Azores dairy farms, technical efficiency is the lowest (66.4%) found when compared to published research (Table 5.1). The above-mentioned literature review reports a mean technical efficiency higher than 88%. Even the number of efficient farms (about 7%) is the second lowest reported value.

The small dimensions (around 25 ha per farms) may explain this low efficiency in the Azores (smaller farms in New Zealand, Canada or Australia). As Cloutier and Rowley (1993) suggest, bigger farms are more efficient. The inefficiency in the Azorean dairy farms seems to be influenced by the great amount of fixed costs spent on agricultural equipment and animal feeding with concentrates.

5.4 Conclusion

The results show that the Azores dairy farms must increase their technical efficiency given that they operate above their resource capacity. It is now necessary to help those small dairy farms in order to make them more efficient.

The efficiency is not directly related to a specific production system. Results show that there are efficient farms in any Azorean farm type: intensive, extensive and intermediate, with different technologies.

Further research must be considered for the DEA concerning relationship between operational research and management science. Cooper (1999) suggests that DEA is a variant of multimodal programming. Using DEA and multicriteria decision-making efficiency must be compared, as suggested by Steward (1996) and Giokas (1997). Finally, it will be interesting to do further research combining different techniques, operational research and statistics.

Acknowledgements This research had financial support of "Fundação para a Ciência e Tecnologia", with the scholarship: PRAXIS XXI/BD/9253/96.

The authors wish to thank Regina Oberschelp de Meneses for editing the English language in this work.

Special thanks to Dr. Tim Coelli for giving the DEAP (DEA software).

References

Alvarez AM, Gonzalez E (1999) Using cross-section data to adjust technical efficiency indexes estimated with panel data. Am J Agric Econ 81:894–901

Arzubi A, Berbel J (2002) Determinación de índices de eficiencia mediante DEA en explotaciones lecheras de Buenos Aires. Invest Ag Prod Sanid Anim 17(1–2):103–123

Avillez F (1991) Estudo de base microeconómica sobre as perspectivas de desenvolvimento da agricultura dos Açores. Instituto Superior Agronomía, Universidade Técnica de Lisboa

Banker R, Charnes A, Cooper WW (1984) Some models for estimating technical and scale inefficiencies in data envelopment analysis. Manag Sci 30(9):1078–1092

Brummer B, Loy J (2000) The technical efficiency impact of farm credit programs – a case study of Northern Germany. J Agric Econ 51(3):405–418

Charnes A, Cooper WW, Rhodes E (1978) Measurement the efficiency of decision making units. Eur J Oper Res 2:429–444

Cloutier L, Rowley R (1993) Relative technical efficiency: data envelopment analysis and Quebec's dairy farms. Can J Agric Econ 41:169–176

Coelli TJ (1995) Recent developments in frontier modelling and efficiency measurement. Aust J Agric Econ 39(3):219–245

Coelli T (1996) A guide to DEAP Version 2.1. A data envelopment analysis computer program. Centre for efficiency and productivity analysis, Department of Econometrics, University of New England, Armidale, Australia

Cooper WW (1999) Operational research/management science – where it's been – where it should be going. J Oper Res Soc 50:3–11

Färe R, Whittaker G (1995) An intermediate input model of dairy production using complex survey data. J Agric Econ 46(2):201–213

Farrell M (1957) The measurement of productive efficiency. J R Stat Soc (Series A) 120(III):253–290

Fraser I, Cordina D (1999) An application of data envelopment analysis to irrigated dairy farms in northern Victoria, Australia. Agric Syst 59:267–282

Giokas D (1997) The Use of goal programming and data envelopment analysis for estimating efficient marginal cost of outputs. J Oper Res Soc 48(3):319–323

Gonzalez Fidalgo E, Alvarez Pinilla A, Arias Sanpedro C (1996) Análisis no paramétrico de Eficiencia en explotaciones lecheras. Investigación Agraria 11(1):173–190

Hallam D, Machado F (1996) Efficiency analysis with panel data – a study of Portuguese dairy farms. Eur Rev Agric Econ 23(1):79–93

Jaforullah M, Whiteman J (1999) Scale efficiency in the New Zealand dairy industry: a non-parametric approach. Aust J Agric Resour Econ 43(4):523–541

Maieta OW (2000) The decomposition of the cost inefficiency into technical and allocative components with panel data of Italian dairy farm. Eur Rev Agric Econ 27(4):473–495

Reinhard S, Thijssen GJ (2000) Nitrogen efficiency of Dutch dairy farms: a shadow cost system approach. Eur Rev Agric Econ 27(2):167–186

Reinhard S, Lovell CAK, Thijssen GJ (1999) Econometrics estimation of technical environmental – an application to Dutch dairy farms. Am J Agric Econ 81:44–60

Reinhard S, Knox Lovell CA, Thijssen GJ (2000) Environmental efficiency with multiple environmentally detrimental variables; estimated with SFA and DEA. Eur J Oper Res 121:287–303

Silva ES (2001) Análisis Multicriterio de la Eficiencia Económica de las Explotaciones Agroganaderas de las Azores (Portugal). Tesis Doctoral. Escuela Técnica Superior de Inginieros de Agrónomos y de Montes, Universidad de Córdoba

Steward TJ (1996) Relationships between data envelopment analysis and multiple criteria decision-analysis. J Oper Res Soc 47(5):654–665

Tauer LW (1988) Productivity of New York dairy farms measured by nonparametric Malmquist indices. J Agric Econ 49(2):234–249

Tauer LW (2001) Efficiency and competitiveness of the small New York dairy farm. J Dairy Sci 84:2573–2576

Weersink A, Turvey CG, Godah A (1990) Decomposition measures of technical efficiency for Ontario, Dairy Farms. Can J Agric Econ 38:439–456

Chapter 6
Animal Grazing System Efficiency

Emiliana Silva, Carlos Santos, and Armando B. Mendes

Abstract This chapter proposes to estimate the technical efficiency in agricultural grazing systems (dairy, beef and mixed) in Azores, in the year 2002. This research used 184 agricultural farms of FADN – Farm Accountancy Data Network. DEA, a non-parametric methodology, was used to estimate efficiency by means of DEAP software.

The results have shown that the average technical efficiency in the dairy grazing system was 63.2% (CRS) and was higher (71.4%) in VRS, and the scale efficiency was about 89.2%. In beef grazing system, the average technical efficiency (CRS) was 69.4%; VRS and the scale efficiency were 82.9 and 84.2%, respectively. In the mixed grazing system, the average technical efficiency (CRS) was 89%, the VRS was higher (99.24%) and the scale efficiency was 89.8%. The mixed system is the *most efficient*, and about half (46.7%) of the farms were efficient. In the dairy grazing system and in the beef systems, only 9.8 and 11.1% were efficient farms. The efficiency is generally higher in mixed systems than in dairy and beef systems.

Keywords Agriculture • Azores • DEA • Efficiency

E. Silva (✉)
CEEAplA, Departamento de Ciências Agrárias, Universidade dos Açores, Rua Capitão João de Ávila, 9700-042 Angra do Heroísmo, Açores, Portugal
e-mail: emiliana@uac.pt

C. Santos
Departamento de Ciências Agrárias, Universidade dos Açores, Rua Capitão João de Ávila, 9700-042 Angra do Heroísmo, Açores, Portugal
e-mail: kamiie@hotmail.com

A.B. Mendes
CEEAplA, Departamento de Matemática, Universidade dos Açores, Rua da Mãe de Deus, 9500-801 Ponta Delgada, Açores, Portugal
e-mail: amendes@uac.pt

A.B. Mendes et al. (eds.), *Efficiency Measures in the Agricultural Sector: With Applications*, DOI 10.1007/978-94-007-5739-4_6,

6.1 Introduction

The main purpose of this chapter is to evaluate the technical efficiency of the three grazing systems in Azores islands (dairy, beef and mixed grazing systems: dairy plus beef), systems previously identified by Silva and Berbel (2006).

The efficiency concept of animal production is an economic and zootechnique perspective, e.g. maximize the production with the lowest number of inputs (land, labour and capital). It is possible to analyse animal efficiency with traditional technical, economic and financial indicators or using parametric (econometric models) or non-parametric techniques (mathematical programming).

It was decided to use DEA (data envelopment analysis), a non-parametric technique to estimate efficiency due to two reasons. First, the objective was to compare each farm with the benchmarking of each grazing system type. Second, because it is possible to use more than one output (milk production and subsidies).

6.2 Efficiency References

The concept of technical efficiency was first introduced by Farrell (1957), but the tool was developed by Charnes et al. (1978). Efficiency was estimated by using the optimization models. DEA make it possible, identifying firms that are standards in their groups. DEA allows the identification of the best practice benchmarking and calculates the relative contribution of each benchmarking (Jaforullah and Whiteman 1999).

To be efficient, a firm must produce in the isoquant. Below this frontier, the firm is inefficient. Farrell (1957) has decomposed the efficiency into (1) technical efficiency and (2) allocative (or price) efficiency. The technical efficiency measures the maximum equiproportion reduction in all inputs that still allow the continued production of the given outputs. This concept means that any point in the curve (named isoquant) is technically efficient.

The allocative efficiency measures the extent of a firm's adaptation to input prices. These two concepts together define economic efficiency. So, economic efficiency only is achieved by one point in the isoquant, the one where the isocost line touches the isoquant. Whenever the isocost is inside or outside the isoquant, some inefficiency occurs. The technical and allocative inefficiency measures the magnitude of consequent loss (what is still possible to produce with those inputs) (Maieta 2000). Similar considerations (the magnitude of loss) are applied to economics efficiency and inefficiency.

The technical efficiency can be measured by constant returns to scale (CRS), the variable returns to scale (VRS) and scale efficiency, oriented to input or output. The CRS assumes that all firms were operating at optimal scale (Charnes et al. 1978). Later, in variable returns to scale (VRS), a change was made to the linear model by incorporating convexity constraints (Banker et al. 1984). This change allowed the decomposition of technical efficiency (or global technical efficiency) into pure technical efficiency and scale efficiency.

If the CRS and VRS are run and operated with the same data and if there is some difference between the firms, that is due to the scale inefficiency. That inefficiency must be calculated from the difference between CRS and VRS. The pure technical efficiency matches with VRS. The scale inefficiency is a result of the fact that the scale level is not optimal (when CRS = 1). The global technical efficiency (CRS) is the product of pure technical efficiency by scale technical efficiency. The use of the VRS specification will allow the calculation of technical efficiency due to these scale efficiency effects.

Researchers have separated the technical efficiency scores obtained from the CRS DEA into two components: (1) due to the scale inefficiency and (2) due to the *pure* technical inefficiency (Coelli 1996). A difference in these two technical scores for a particular decision-making unit indicates that the decision-making unit has scale inefficiency. That scale inefficiency can be calculated from the difference between VRS and CRS technical efficiency scores.

DEA was used in various researches with some applications on dairy farms such as Silva et al. (2004), Venâncio (2003), Marote and Silva (2002), Arzubi and Berbel (2001), Reinhard and Thijssen (2000), Jaforullah and Whiteman (1999), Fraser and Cordina (1999), Alvarez and Gonzalez (1999), Tauer (1998), Rico (1996), González Fidalgo et al. (1996), Färe and Whittaker (1995) and Cloutier and Rowley (1993).

Other recent works used a data set of several economic variables from Azores dairy farms. Noncheva et al. (2009a, b) developed a case study for illustrating canonical correlation analysis in variable aggregation. Mendes et al. (2010) used canonical correlation analysis in variable selection. The same approach was used by Noncheva et al. (2009c) for the development of a software framework for measuring efficiency by creating a user-friendly interface for the packages of DEA in R software.

6.3 Material and Methods

This work used 184 farms of FADN-Azores for the year 2002. From these, 133 were dairy farms, 36 beef farms and 15 mixed farms (dairy and beef).

In this work, the most important costs were selected: (1) the intermediate costs (feeding; fertilizers; seeds; hired machinery; machinery and buildings repairs; fuel and lubricants; other animal cash expenses such as medicine, artificial insemination and hygienic products; and other costs) and (2) fixed costs (rent paid and depreciation of machinery and building). The two outputs were animal production and subsidies.

A non-parametric model, data envelopment analysis, DEA, was used to estimate the efficiency farms and uses a linear programming to construct an efficient frontier that provides a means by which all farms can be assessed in terms of relative efficiency (Fraser and Cordina 1999).

DEA was used mainly for benchmarking performance and allows the measurement of technical efficiency of individual farms according to their all quantifiable variables. Efficiency is a relative concept because benchmarking is a procedure for improving performance by identifying best practice, measuring performance against best practice and benchmarking partnerships between best practice and non-best practice enterprises, and that will allow eliminating the less efficient practice (Jaforullah and Whiteman 1999).

DEA allows input and output orientation models, each one a constant or variable returns to scale.

One of the advantages of DEA is the possibility of dealing with multiple inputs and outputs, and this characteristic was explored in this research. Three models were considered (dairy, beef and mixed) to estimate the efficiency. There were two outputs (animal product and subsidies) and three aggregate inputs.

The outputs were milk and beef production (€) (Y_1) and subsides (€) (Y_2).

The inputs were agricultural areas (SAU) (X_1) (ha), annual work unit (AWU) (X_2) (h) and total costs (X_3) (machinery and building repairs (€), fuel and lubricants (€), animal feeding (€), other animal cash expenses (€), fertilizers and herbicide (€), rent paid (€) and equipment depreciation (€)).

The DEAP Version 2.1 developed by Coelli (1996) was used, VRS and CRS input oriented, because it was expected that the input could influence the inefficiency of farms.

The most important costs were in feeding the animals and fertilizers. Feeding costs were the most important in dairy (45.8%) and mixed farms (33.3%). In beef farms, feeding costs were lower (25.8%). In the 133 dairy farms, the feeding costs were nearly 41.5%. The importance of both food and fertilizer costs was confirmed by previous researches: Venâncio (2003), Marote and Silva (2002), Silva (2001), and Silva and Berbel (2004).

Fertilizer costs were more important in beef farms (25.8%) than those in dairy or mixed farms (less than 14%). For all farms, the costs were nearly 15.13%. In the beef grazing system, there was a replacement from green feeding (fertilizers) to dry feeding.

The importance of intermediate costs was for other animal cash expenses (9.9%), fuel and lubricants (9.5%), machinery and building repairs (8.4%), machinery (6%), other costs (3.8%) and, finally, seeds (2.7%).

The machinery depreciation costs and rent payments were the most representative fixed costs in the three grazing systems.

The depreciation of machinery was similar in beef and mixed grazing systems, 58.5 and 54.8%, respectively, and lower in milk grazing systems (44.5%). The machinery depreciation was nearly 46.6% of the total fixed costs in all FADN farms.

The rent payments were more noticeable in milk grazing systems (33%) and less representative in beef grazing systems (29.1%). The other fixed costs (wage, taxes, insurance) were not so noticeable (less than 8%).

It must be detached from the fact that most of the FADN Azorean farms are family businesses, which means that the farm work is mainly familiar and not effectively paid.

The most important outputs were livestock, milk and dairy production and subsidies. The subsidies were bigger in beef grazing systems (approximately 60%)

due to the European agricultural policy (subsidy for the suckler cow sector, beef special premium scheme, slaughter premium scheme and enlargement). In milk grazing systems, the subsidies were less important (about 20% of total output).

The vegetable production was not representative (less than 2.5% of the total production) and was a complement of animal production. In the dairy grazing system, there was only 0.65% of vegetable production, and the beef grazing system was a bit higher (1.9%).

6.4 Discussion

The average size of farms was 42.3 ha. The farms were classified according to the European classification (OTE). In dairy farms (OTE 4.1), the average size was 28 ha. The average size of the beef farms (OTE 4.2) was 65 ha, and the largest one had 801.9 ha. In the mixed farms (OTE 4.3.), the average size of farms was 34 ha. The dairy farms are typically less utilized agricultural areas, and beef farms are the biggest ones.

The intensification of the system was measured by the stocking density. The average stocking density in FADN-Azores was 1.05 bovines per hectare and 1.91, 0.43 and 1.27 in dairy, beef and mixed grazing systems, respectively. The dairy farms had a bigger level of intensification (more stocking density), and the beef farms had the biggest extensification (less stocking density).

The annual work unit (AWU) is about 1.28 for the FADN-Azores. In the dairy, beef and mixed farms, it was 1.37, 0.97 and 1.19, respectively. The dairy grazing systems need more AWU as a consequence of the higher intensity.

The net income per animal in dairy farms was nearly 426€, in the beef farms it was 360€ and in the mixed farms it reached 201€. The gross income per animal was 727€ in dairy farms, 528€ in beef farms and 511€ in mixed farms. The subsidies per animal were higher in beef farms (about 389€) and lower in dairy farms (170€). In the mixed farms, it was nearly 200€ per animal. The total cost was higher in dairy farms (518€ per animal) and lower in beef farms (223€ per animal).

In dairy farms, the average technical efficiency was 63.2% to CRS, 71.4% to VRS and 89.2% when compared to the scale efficiency.

The standard deviation is low and about 0.133–0.189 (Fig. 6.1). The lowest value of efficiency (CRS) was reached in the animal production systems, which is significant and means that this farm could produce the same using only 15.2% of inputs.

In the beef grazing system, the average technical efficiency (CRS) was about 69.4%, the VRS was 82.9% and 84.2% of the scale efficiency (Fig. 6.2). The standard deviation was also low, from 0.171 (CRS) to 0.199 (VRS). The lowest value of efficiency (0.347) is slightly superior to dairy farms, which has the same meaningalready described for the milk grazing system.

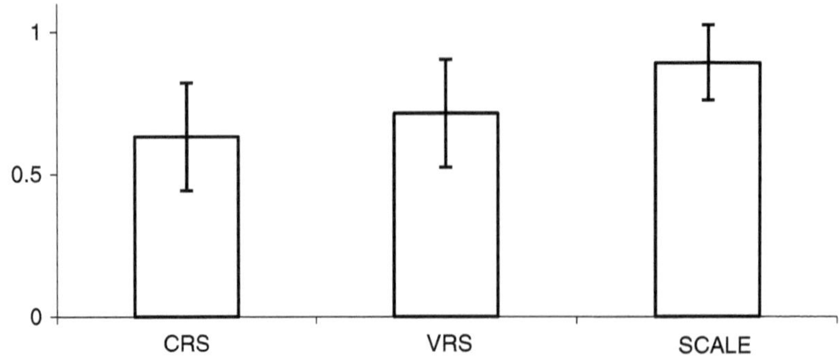

Fig. 6.1 Average technical efficiency (*CRS*, *VRS* and *SCALE* – scale efficiency) in milk grazing system

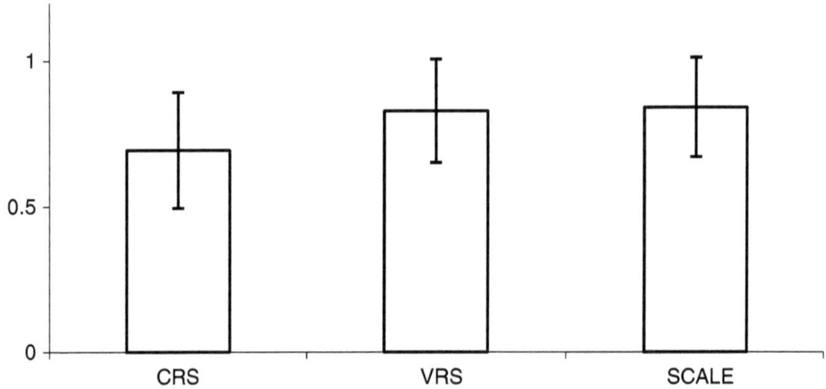

Fig. 6.2 Average technical efficiency (*CRS*, *VRS* and scale efficiency) in beef grazing systems

The average technical efficiency in mixed grazing system to CRS was nearly 89%, to VRS was approximately 99.2% and the scale efficiency was 89.8% (Fig. 6.3). The standard deviation is still small, in CRS or VRS, ranging from 0.023 to 0.148. The lowest value of efficiency in mixed system was 50.4% meaning that it is technically possible to produce the same amount of product with just half of the inputs.

In the FADN-Azores survey, the mixed grazing system (although it was not the most noticeable) is the *most efficient* with the highest values of efficiency (CRS, VRS and scale).

In Table 6.1, the dairy farms' frequency is bigger, about 78% (104 farms in CRS), in the interval efficiency lower than 0.750. In this system, we can find the lowest value for efficiency (0.152). There are (9.8%) farms that are efficient, with the value of 1 in CRS and in the scale efficiency. When the VRS efficiency is considered, the number of efficient farms increases to 22 (16.5% efficient farms).

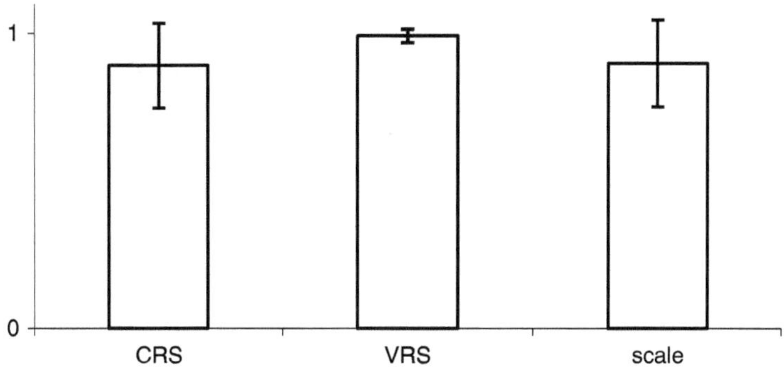

Fig. 6.3 Average technical efficiency (*CRS*, *VRS* and scale efficiency) in mixed grazing systems

Table 6.1 Technical efficiency intervals for milk grazing system, in 2002

		Dairy	
Efficiency interval	CRS	VRS	Scale
<0.750	104	78	17
0.750–0.949	14	33	60
0.950–0.999	2	0	43
1	13	22	13
Total	133	133	133

Table 6.2 Technical efficiency interval for beef grazing system in 2002

		Beef	
Efficiency interval	CRS	VRS	Scale
<0.750	24	11	10
0.750–0.949	6	9	13
0.950–0.999	2	5	9
1	4	11	4
Total	36	36	36

In the beef production system, most of the farms, approximately 66% (24 farms in the CRS), are operating in the interval lower than 0.750 (Table 6.2). The number of efficient farms is lower than in the other grazing systems. There are only 4 (11.1%) efficient farms in CRS and scale efficiency. This figure increases for VRS (30.5% of farms are efficient).

In the mixed production system, there were about 50% efficient farms: 7 efficient farms out of 15 (CRS). There were 6 farms operating in the efficiency interval from 0.750 to 0.949 (Table 6.3). In the VRS, most parts (86.6%) of the farms are efficient (VRS equal the unity), but farms still persist in the interval of efficiency 0.750–0.949.

Table 6.3 Technical efficiency interval for mixed grazing system in 2002

	Mixed		
Efficiency interval	CRS	VRS	Scale
<0.750	2	0	2
0.750–0.949	6	2	4
0.950–0.999	0	0	1
1	7	13	8
Total	15	15	15

6.5 Conclusions

The dairy farms' efficiency now has a lower value when compared with the previous years (Silva 2001; Marote and Silva 2002; Venâncio 2003).

In Azores, different grazing systems (milk, beef and mixed) have different values of technical efficiency. This is a consequence of competing by differentiating the grazing systems and also because of the different magnitudes of the three systems. In fact, the technical efficiency is relative to a standard firm, each being different from system to system and so not really comparable across systems.

In FADN-Azores, the average technical efficiency (CRS) was approximately 55.8%. In the mixed grazing system, it was 89%, the beef grazing system being 69.4% and the milk being 63.2%. The heterogeneity of grazing systems increased the inefficiency: in each grazing system, some farms were efficient but became inefficient when analysed all together.

Although using parametric methods, Venâncio (2003) in Azores had a similar result: the mixed system was the most efficient grazing system.

Some farms had a lower efficiency due to the excess of feeding, depreciation machinery and amount of fertilizers (the most representative costs).

As a consequence, the increasing technical efficiency of Azorean farms needs a rational use of some inputs: feeding, depreciation, fertilizers and rent paid.

References

Alvarez A, Gonzalez E (1999) Using cross-section data to adjust technical efficiency indexes estimated with panel data. Am J Econ 81:894–901

Arzubi A, Berbel J (2001) Determinación de la Eficiencia usando DEA en Explotaciones Lecheras en Argentina. IV Congreso de la Asociación Española de Economía Agraria. "Economia Agraria y Recursos naturales: Nuevos Enfoques y Perspectivas". Pamplona España

Banker R, Charnes A, Cooper W (1984) Models for estimation of technical and scales inefficiency in data envelopment analysis. Manag Sci 30:1078–1092

Charnes A, Cooper W, Rhodes E (1978) Measurement the efficiency of decision making units. Eur J Oper Res 2:429–444

Cloutier L, Rowley R (1993) Relative technical efficiency: data envelopment analysis and Quebec's dairy farms. Can J Agric Econ 41:169–176

Coelli T (1996) A guide to DEAP Versión 2.1: a data envelopment analysis computer program. Centre for Efficiency and Productivity Analysis, Department of Econometrics, University of New England, Armidale, Australia

Färe R, Whittaker G (1995) An intermediate input model of dairy production using complex survey data. J Agric Econ 46(2):201–213

Farrell M (1957) The measurement of productivity efficiency. J R Stat Soc (Series A) 120(III):253–290

Fraser I, Cordina D (1999) An application of data envelopment analysis to irrigated dairy farms in Northern Victoria, Australia. Agric Syst 59:267–282

González Fidalgo E, Alvarrez Pinilla A, Arias Sampedro C (1996) Análisis no parametrico de eficiencia en Explotaciones lecheras. Investigación Agraria 11(1):173–190

Jaforullah M, Whiteman J (1999) Scale efficiency in the New Zealand dairy: a non-parametric approach. Aust J Agric Resour Econ Farms 43(4):523–541

Maieta O (2000) The decomposition of cost inefficiency into technical and allocative components with panel data of Italian dairy farms. Eur Rev Agric Econ 27(4):473–495

Marote E, Silva E (2002) Análise Dinâmica das Explorações Leiteiras da Ilha Terceira. Actas do XII Congresso de Zootecnia

Mendes A, Silva E, Noncheva V, Santos J (2010) Terceira *farms' efficiency measurement using DEA and CCA*. In: Silva E et al (eds) Proceedings do congresso da APDEA e Gestão e Conservação da Natureza. Univ. dos Açores, Ponta Delgada

Noncheva V, Mendes A, Silva E, Santos J (2009a) Canonical correlation analysis in variable aggregation in DEA. In: Póvoa A, Salema M (eds) Livro de Actas, Congresso IO2009, FCT-UNL, pp 115–122

Noncheva V, Mendes A, Silva E (2009b) An approach to variable aggregation in efficiency analysis. In: Classification, forecasting, data mining, international book series information science & computing. Int J Inf Technol Knowl 3(8):97–104

Noncheva V, Mendes A, Silva E (2009c) A *software framework for measuring efficiency, book of abstracts,UseR Conference*, Rennes, France, 8–10 July , pp 144–144

Reinhard S, Thijssen G (2000) Nitrogen efficiency of Dutch dairy farms: a shadow cost system approach. Eur Rev Agric Econ 27(2):167–186

Rico E (1996) Análisis Dinámico de la Estructura Economico-Financiera y de la Eficiencia de las Almazaras Cooperativas en la Provincia de Jaén Durante el Período 1975–1993. Tesis doctoral, ETSIAM, Universidad de Córdoba

Silva E (2001) Análisis de la Eficiencia Económica de las Explotaciones Agroganaderas de las Azores (Portugal). Tesis doctoral, Universidad de Córdoba

Silva E, Berbel J (2004) A decision support model for dairy farms. In: Schiefer G, Rickert U (eds) Quality assurance, risk management and environmental control in agriculture and food supply networks, Proceedings of the 82nd EAAE, vol B, 14–16 May, pp 583–592. (ISBN 3-935887-65-4)

Silva E, Berbel J (2006) An Azorean farms typology. New Medit VI(1):51–54

Silva E, Berbel J, Arzubi A (2004) (DEA) in Azores dairy farms. New Medit III(3):39–43

Tauer LW (1998) Productivy of New York dairy farms measured by nonparametrics malmquist indices. J Agric Econ 49(2):234–249

Venâncio F (2003) Eficiência Técnica das Explorações da Ilha do Faial: o método da fronteira estocástica, Relatório de Estágio em Engenharia Zootécnica, Departamento de Ciências Agrárias, Universidade dos Açores, Angra do Heroísmo

Chapter 7
Technical Efficiency of the Spanish Dairy Processing Industry: Do Size and Exporting Matter?

Magdalena Kapelko and Alfons Oude Lansink

Abstract This chapter uses DEA to measure the technical efficiency of a sample of Spanish dairy processing firms over the period 2001–2009. Differences in technical efficiency between firms of different sizes and between firms that operated in international markets versus those that were not are tested. The results show that larger dairy processing firms were, on average, more efficient than smaller ones. Furthermore, within the groups of small and large firms, those firms that are exporting were more efficient than firms that do not export. The distribution of technical efficiency, of the small- and micro-sized exporters, stochastically dominates the distribution of the non-exporters in most years.

Keywords Technical efficiency • Data envelopment analysis • Dairy processing industry

7.1 Introduction

Over the past decade, the European Union (EU) introduced a number of measures that aim at liberalizing the EU dairy market (Boel 2008). Common Agricultural Policy (CAP) reforms have resulted in a lower budgetary support for the production of bulk dairy products (milk powder and butter). Also, the EU liberalizes its dairy market through the abolishment of the dairy quota system in 2015. The liberalization comes in a period of ongoing globalization, in which competition from major

M. Kapelko (✉)
Institute of Applied Mathematics, Wroclaw University of Economics, Wroclaw, Poland
e-mail: magdalena.kapelko@ue.wroc.pl

A. Oude Lansink
Business Economics Group, Wageningen University, Wageningen, The Netherlands
e-mail: alfons.oudelansink@wur.nl

A.B. Mendes et al. (eds.), *Efficiency Measures in the Agricultural Sector: With Applications*, DOI 10.1007/978-94-007-5739-4_7,

exporters such as Australia and South America increases, and new opportunities arise in emerging markets such as China, India and Russia (Kleibeuker 2007). At the same time, dairy markets were increasingly volatile, causing increasing risks for producers and other stakeholders in the dairy supply chain. Also, consumer preferences are changing towards higher value-added dairy products requiring investments in product innovation (Devlieghere et al. 2004). These developments require an improvement of the competitiveness of the EU dairy sector, in which the dairy processing industry plays a key role of processor and marketer of raw milk produced by dairy farmers.

This study focuses on dairy processing firms in Spain. The dairy processing industry in Spain is dominated by small- and medium-sized companies, of which some are active in international markets. In the currently rapidly changing environment with increasing globalization and liberalization, it is likely that the poorly performing, less competitive dairy processing firms will quit the sector. This raises questions like the following: What is the current performance of Spanish dairy processing firms? What is the impact of firm size and participation in international markets on the performance of Spanish dairy processing firms?

The literature has used a variety of methods to determine the performance of dairy processing companies. One stream of literature uses financial ratios to measure several aspects of the financial performance of dairy processing firms and compare them with other firms (see Soboh et al. (2009) for an overview). Another stream of research analyses the technical efficiency using either data envelopment analysis (DEA) or stochastic frontier analysis (SFA) (e.g. Singh et al. (2001) for Indian dairy processing firms, Doucouliagos and Hone (2001) for Australian dairy processing firms and at last, Boyle (2004) for Irish dairy processing firms). Soboh et al. (2012) used different DEA models to estimate and compare the technical efficiency of EU dairy processing cooperatives and investor-owned firms (IOF) in Germany, Ireland, France, Denmark, the Netherlands and Belgium. They conclude that cooperatives were on average less efficient than IOFs when using a traditional DEA model but more efficient than IOFs when using a DEA model that is orientated from the role of cooperatives as a company that serves the objectives of its members (dairy farmers).

The empirical literature found strong evidence for a positive association between participation in international markets and performance. This relation was theoretically pointed in two distinct ways (Delgado et al. 2002). First, it was hypothesized that only the more competitive firms can participate in international markets since only these firms were able to bear the high costs of entering foreign markets. Second, it was expected that firms that are exposed to competition in international markets get more incentives to operate efficiently than those that only sell in protected local markets. Empirical evidence from Chen and Tang (1987), Aw and Hwang (1995), Bernard and Jensen (1995), Jensen and Wagner (1997), Aw et al. (1997, 2000), Clerides et al. (1998) and Delgado et al. (2002) confirmed that export-oriented firms are more efficient than non-exporter ones.

The relationship between size and performance can be argued in different ways. Large firms are in a better position to use economies of scale (Majumdar 1997); hence, size is often expected to have a positive impact on performance. However, size is also correlated with market power which may be the source of

x-inefficiencies, leading to a more inferior performance (Leibenstein 1976). Also, specifically for dairy processing firms, Soboh et al. (2012) argue that cooperatives (which frequently occur in dairy processing) suffer from the obligation to process all their members' milk. This obligation may force cooperatives to operate at a too large scale, hence causing a lower performance.

In the light of the foregoing, the objectives of this chapter were to measure the technical efficiency of dairy processing firms in Spain and to test for the significance of differences in technical efficiency between dairy processing firms of different sizes and firms that are active in export markets versus those that are not. A sample of dairy processing firms from Spain over the period 2001–2009 is the focus of the empirical application.

The remainder of this chapter is organized as follows. The next section discusses the data envelopment analysis methodology used in this chapter and presents the approach to test various hypotheses. This is followed by a description of the data, the presentation of the results and the conclusions.

7.2 Methodology

This section explains the foundations of data envelopment analysis (DEA) model used to compute the technical efficiency scores of analysed firms. Data envelopment analysis (Charnes et al. 1978; Banker et al. 1984) involves the application of linear programming techniques to observed inputs used and outputs produced by decision-making units (DMUs) under analysis to obtain efficiency measures. DMUs are, for example, firms, departments, countries or, in this study, dairy processing firms. In the next step, DEA constructs a production frontier reflecting the best practices. Each firm's technical efficiency is then measured relative to this frontier.

Applying DEA requires choosing the orientation of the model and returns to scale imposed on the technology. In particular, technical efficiency can be assessed in input and output orientation: the former model aims at minimizing inputs while keeping outputs constant, while the latter maximizes outputs at the originally specified input levels. There exists also a hyperbolic orientation which simultaneously focuses on increasing outputs and minimizing inputs. This research uses an input-oriented model, and the dairy processing firms are viewed as cost minimizers. As such, it assumes that the firms' managers can exercise control over the inputs, while outputs are defined by demand and limited only by resources and capacities. Returns to scale measure the change in output levels due to a 1% increase of all inputs. The original DEA model proposed by Charnes et al. (1978) assumes constant returns to scale (CRS), while the variable returns to scale model (VRS) is developed by Banker et al. (1984). Constant returns to scale implies that an increase in all inputs results in a corresponding proportional increase in outputs. Variable returns to scale indicate that an increase in the input levels does not necessarily result in a proportional increase in output levels, that is, the output levels can increase more than proportionally (increasing returns to scale – IRS) or less than proportionally (decreasing returns to scale – DRS) as the increase in all inputs. In general, the VRS assumption allows for a comparison of firms with similar size (Coelli et al. 2005). In this study, we choose VRS due to the fact that our data set includes firms that largely differ in size.

The mathematical formulation of the VRS input-oriented model goes as follows. Suppose we have n DMUs to be evaluated, and each of them consumes varying amounts of m different inputs to produce s different outputs. DMU_k consumes quantity $X_k = \{x_{ik}\}$ of inputs $i = \{1, 2, \ldots, m\}$ and produces quantity $Y_k = \{y_{jk}\}$ of outputs $j = \{1, 2, \ldots, s\}$. The model evaluates the efficiency score of each observed DMU called DMU_o relative to other DMUs. The linear programming model is described below:

$$
\begin{aligned}
&\text{Min } \lambda \\
&\text{subject to } \sum_{k=1}^{n} x_{ik} z_k \leq \lambda x_{io}, \quad i = \{1, 2, \ldots, m\} \\
&\sum_{k=1}^{n} y_{jk} z_k \geq y_{jo}, \quad j = \{1, 2, \ldots, s\} \\
&\sum_{k=1}^{n} z_k = 1 \\
&z_k \geq 0
\end{aligned}
\tag{7.1}
$$

where

λ is the input-oriented efficiency coefficient.
x_{ik} stands for the quantity of input $i = \{1, 2, \ldots, m\}$ used by DMU_k ($k = 1, \ldots, n$).
y_{jk} stands for the quantity of output $j = \{1, 2, \ldots, s\}$ produced by DMU_k.
x_{io} represents the quantity of input i used by the observed unit under analysis DMU_o.
y_{jo} represents the quantity of output j produced by the observed unit under analysis DMU_o.
z_k symbolizes the activity levels associated with inputs and outputs of DMU_k.

The computation of efficiency scores involves solving one linear program for each DMU.

In order to provide the statistical significance to computed DEA indexes, bootstrap techniques can be used. Bootstrapping is defined as a repeated simulation of the data-generating process through resampling and applying the original estimator to each simulated sample so that resulting estimates imitate the original unknown sampling distribution of the estimators of interest (Simar and Wilson 1998). To introduce the bootstrap procedure mathematically, we denote $\chi = \{(\mathbf{x}_k, \mathbf{y}_k), k = 1, \ldots, n\}$ as an original sample of n DMUs for which bootstrap should be estimated. The algorithm can be summarized in the following steps (Simar and Wilson 1998, 2000):

1. The computation of the efficiency scores $\hat{\delta}_k$ for each decision-making unit $k = 1, \ldots, n$ by solving the linear programming model described by (7.1).
2. Using kernel density estimation and reflection method (smooth bootstrap), the generation of the pseudo sample $\chi^* = \{(\mathbf{x}_k^*, \mathbf{y}_k^*), k = 1, \ldots, n\}$ to form the reference bootstrap technology; applying smooth bootstrap that is drawing with replacement from the density of original efficiency scores (using kernel smoothing and reflection method) yields consistent estimators; when this density is already estimated, drawing a pseudo sample involves solving the additional linear programs.
3. The computation of the bootstrap estimation of efficiency $\hat{\delta}_{kb}^*$ of $\hat{\delta}_k$ for each $k = 1, \ldots, n$ by applying the original estimators to the pseudo data set derived from step 2.
4. The repetition of steps 2 and 3 B times in order to obtain a set of estimates $\left\{\hat{\delta}_{kb}^*, b = 1, \ldots, B\right\}$.

After the bootstrap values are calculated, the efficiency measures can be corrected for bias, and confidence intervals of efficiency scores can be computed.

Bootstrap bias values for the original estimator $\hat{\delta}_k$ can be estimated from the following equation:

$$\text{bia}\hat{s}_B\left(\hat{\delta}_k\right) = B^{-1}\sum_{b=1}^{B}\hat{\delta}_{kb}^* - \hat{\delta}_k$$

and the bias-corrected estimator of efficiency can be finally computed as:

$$\hat{\hat{\delta}}_k = \hat{\delta}_k - \text{bia}\hat{s}_B\left(\hat{\delta}_k\right) = 2\hat{\delta}_k - B^{-1}\sum_{b=1}^{B}\hat{\delta}_{kb}^*$$

The computation of confidence intervals for efficiency scores involves the following steps:

- Sort the values $\left(\hat{\delta}_{kb}^* - \hat{\delta}_k\right)$ for $b = 1, \ldots, B$ and delete $(\alpha/2 \times 100)$ percent of the elements at either end of this sorted array.
- Set $-\hat{b}_\alpha^*$ and $-\hat{a}_\alpha^*\left(\hat{a}_\alpha^* \leq \hat{b}_\alpha^*\right)$, equal to the endpoints of the resulting array, then the estimated $(1-\alpha)$ percent confidence interval is formulated as $\hat{\delta}_k + a_\alpha^* \leq \delta_k \leq \hat{\delta}_k + b_\alpha^*$.

In this chapter, the bootstrap procedure is implemented using FEAR 1.13 package with $B = 2{,}000$ bootstrap replications. FEAR is freely available software for frontier efficiency analysis with *R*. It is written by Wilson (2008).

This study also tests whether the distribution of efficiency scores of firms that export and firms that do not export differ using the stochastic dominance criterion. The test of stochastic dominance is more general than the Wilcoxon as it tests if the entire distribution is different. It is designed and implemented for efficiency studies by Delgado et al. (2002). Stochastic dominance refers to differences between a pair of distributions, which are characterized by their cumulative distribution functions. Formally, assume that we have two distributions *A* and *B* with cumulative distribution functions *F* and *G*, respectively. First-order stochastic dominance of *A* relative to *B* is defined by $F(x) - G(x) \leq 0$ for any argument $x \in R$ (Delgado et al. 2002). We need to test the following hypotheses:

1. H_0:F(x) = G(x) for all x ∈ R versus
   H_1: F(x) ≠ G(x) for at least one value of x, and
2. H_0:F(x) – G(x) ≤ 0 for all ∈ R versus
   H_1:F(x) – G(x) > 0 for at least one value of x.

In particular, to show that one distribution stochastically dominates the other, we have to demonstrate that the first hypothesis is rejected (the distributions are significantly different), while the second hypothesis cannot be rejected (one distribution is to the right of the other). So, if we assume that the efficiency distributions of exporters stochastically dominate the efficiency distributions of non-exporters, this implies that exporting firms tend to be more efficient than non-exporting firms. To test the hypotheses, the Kolmogorov-Smirnov two-sided and one-sided tests are used (Conover 1971). Because the application of this test requires the independence of observations, the test statistics need to be calculated separately for each analysed year. The test is implemented using STATA 10.0.

7.3 Information About Sample and Variables

The data used in this study comes from the SABI (Sistema de Análisis de Balances Ibéricos) database. The SABI contains financial accounts of a large number of Spanish companies, classified according to the NACE Rev. 2 code, which is a statistical classification of economic activities used in the European Union. To focus on the dairy processing industry, we considered the four-digit code 10.51, which refers to the operation of dairies and cheese making and belongs to the general code 10.5 – manufacture of dairy products. The analysed firms produce a number of different products: liquid milk, butter, yoghurt and cheese. Initially, 705 firms are extracted from the database. Then some companies that do not provide all the information that is necessary to compute efficiency scores were filtered, and outliers were determined using ratios of output to input: an observation was defined as an outlier and removed from the sample if the ratio of output over any of the three inputs was outside the interval of the median plus and minus two standard deviations. The final sample consists of unbalanced panel of 3,509 observations of dairy processing firms that operated in Spain from 2000 to 2009 (Table 7.2).

As it was explained before, to compute DEA efficiency, one needs the information on firms' inputs and outputs. For this purpose, we selected the following

Table 7.1 Descriptive statistics of the data (2000–2009)

Variable	Min	Max	Mean	Std. dev.
Inputs				
Fixed assets	0.848	229,413.800	3,447.751	16,920.670
Employee cost	0.955	105,316.800	873.990	4,633.483
Material cost	1.691	453,163.400	6,984.567	30,199.800
Output				
Turnover	3.645	1,090,795.000	10,011.660	47,970.370

Note: The values of input and outputs are presented in thousands of euros, constant prices from 2000

accounting variables: turnover as output and employee cost, material cost and fixed assets as inputs, following Soboh et al. (2012) for the European dairy processing industry research. The inputs and outputs were expressed in thousands of euros of the year 2000 by deflating the monetary values with corresponding price indexes. Turnover was deflated using industrial price index for output for manufacture of dairy products, fixed assets through industrial price index for capital goods and material cost using industrial price index for consumer nondurables, while labour cost index in manufacturing was applied to deflate employee costs. All indices were obtained from the Spanish Statistical Office (2012).

Table 7.1 provides the basic descriptive statistics of the variables for the 2000–2009 time period.

Table 7.1 shows that the average dairy company in our sample is relatively small in terms of the EU size classification, with a mean turnover of ten million euros. On the other hand, the standard deviations are relatively high, showing that the firms in our sample differ considerably in terms of their inputs and output.

7.4 Results and Discussion

Table 7.2 summarizes the evolution of efficiency scores, bias-corrected efficiency scores and confidence intervals for bias-corrected efficiency scores during the period 2000–2009.

Table 7.2 shows that during the period under investigation, on average dairy processing firms in this sample had relatively high levels of efficiency of 0.723 with a standard deviation of 16.4%. However, the fraction of firms that was classified as efficient is rather low (approximately 9.4%). The analysis of the bias-corrected efficiencies shows that mean efficiency of the sample decreases to 0.664 with a standard deviation of 13.9%. This value indicates that there was a scope for efficiency improvement for firms by reducing the inputs. The confidence intervals show that the values of bias-corrected efficiency scores were contained in the set between 0.621 and 0.714. Furthermore, Table 7.2 reveals fluctuations of efficiencies in the period analysed as well as a reduction in efficiency between 2000 and 2009,

Table 7.2 Temporal evolution of technical efficiency

Year	No of firms	Efficiency score	Fraction of efficient firms	Bias-corrected efficiency score	Confidence interval	
					Lower bound	Upper bound
2000	264	0.798 (0.139)	0.14	0.746 (0.118)	0.701 (0.107)	0.792 (0.137)
2001	319	0.701 (0.176)	0.091	0.638 (0.146)	0.592 (0.131)	0.691 (0.173)
2002	362	0.697 (0.169)	0.094	0.632 (0.137)	0.586 (0.122)	0.687 (0.166)
2003	367	0.777 (0.125)	0.095	0.729 (0.101)	0.689 (0.093)	0.772 (0.124)
2004	372	0.725 (0.174)	0.113	0.667 (0.145)	0.622 (0.132)	0.717 (0.171)
2005	372	0.744 (0.146)	0.097	0.687 (0.114)	0.644 (0.100)	0.737 (0.143)
2006	372	0.722 (0.166)	0.097	0.661 (0.137)	0.617 (0.163)	0.712 (0.125)
2007	345	0.714 (0.161)	0.07	0.657 (0.138)	0.617 (0.129)	0.704 (0.158)
2008	380	0.731 (0.148)	0.095	0.677 (0.122)	0.635 (0.113)	0.724 (0.145)
2009	356	0.633 (0.177)	0.059	0.563 (0.142)	0.517 (0.125)	0.620 (0.172)
Mean 2000–2009	3509	0.723 (0.164)	0.094	0.664 (0.139)	0.621 (0.128)	0.714 (0.162)

The values presented in brackets are standard deviations

especially considerable in 2009. In this year, milk prices dropped to historically low levels in Spain. Delivery of milk to the dairy processing firms declined as well as the production of value-added products like cheese and basic products like butter and milk powder (Eurostat 2012), which might explain the efficiency decline. The efficiency scores in this study were higher than the score of 0.56 found for the entire Spanish food and beverages by Martin-Marcos and Suarez-Galvez (2000). The scores were also higher than these found for dairy processing firms in Belgium, Denmark, France, Germany, Ireland and the Netherlands by Soboh et al. (2012). This suggests that dairy processing firms were on average closer to their own frontier than the firms in the six EU countries in the study of Soboh et al. (2012).

Given that variation in efficiency scores in the sample may be due to differences in their size, the comparison of the bias-corrected indices across four size intervals based on the number of employees and turnover according to the EU definition of micro, small, medium and large firms was made. Following this definition, the category of micro/small/medium firms was made up of enterprises which employ less than 10/50/250 employees and which have an annual turnover not exceeding 2/10/50 million euros, respectively. The firms with more than 250 employees and an annual turnover exceeding 50 million euros are defined as large. These results are presented in Table 7.3.

To test the statistical significance of the mean differences across size intervals, the Kruskal-Wallis test was used. This test is a generalization of the Mann-Whitney test. Comparing to the later, it deals with more than two groups of observations that need to be contrasted. The further discussion of the application of this test in efficiency studies is discussed in Sueyoshi and Aoki (2001). The test is implemented using STATA 10.0. Efficiency scores differ markedly across size groups which reflect the fact that efficiency depends on the firm's size. The group of large firms had the highest values of efficiency scores, with an average bias-corrected efficiency

Table 7.3 Bias-corrected efficiency scores by firms' sizes (2000–2009)

Size class	No of firms	Bias-corrected efficiency score
Large	158	0.763 (0.073)
Medium	337	0.745 (0.105)
Small	700	0.716 (0.107)
Micro	2,314	0.630 (0.142)
Kruskal-Wallis test		463.106*

The values presented in brackets are standard deviations
*Statistically significant differences at 1% level

of 0.763 for the 2000–2009 time period. In contrast, micro firms had the lowest values of efficiency scores, while the values for medium and small locate in between. Therefore, in the sample analysed, the bias-corrected efficiency scores increase with the size of firms.

Furthermore, a test of stochastic dominance was performed to evaluate if firms participating in international markets were more efficient than firms serving domestic markets only. In particular, it was tested if distributions of bias-corrected efficiency scores of exporting firms were different and dominated these of non-exporting firms. Table 7.4 presents the results of the test for stochastic dominance of exporting firms over non-exporting firms.

As indicated in Table 7.4, the null hypothesis of equal distributions of exporters versus non-exporters was rejected for all years, mostly at the 1% critical level. Also, the null hypothesis that exporters had higher efficiency than non-exporting firms cannot be rejected. Therefore, dairy processing firms participating in international markets had higher efficiency during the period analysed. This might be due to the fact that these firms were exposed to more competition, which tends to eliminate inefficiency. Additionally, Fig. 7.1 below provides a graphical representation of the cumulative distribution functions of bias-corrected efficiencies for exporters and non-exporters for the entire time period 2000–2009. One can visually assess that the efficiency distribution of exporters lies to the right of the distribution of non-exporters, which indicates higher efficiency levels for exporting companies.

After obtaining conclusions about efficiency differences between exporters and non-exporters in the entire population of firms, it was analysed whether the results for the whole sample were similar to the groups of micro and small firms and medium and large firms. Although previously in this chapter, the four size classes were treated separately, at this stage, we decided to group small firms with micro and medium with large. This was done because the sample was too small when exporters and non-exporters were distinguished within one size class, especially for the case of large firms. Table 7.5 shows the stochastic dominance test statistics for the differences between exporters and non-exporters by size class.

For the group of micro and small firms, the null hypothesis of equality between distributions of exporters versus non-exporters was rejected at the 5% level for all years, except for 2005, 2007, 2008 and 2009. The null hypothesis that micro and small exporters have a larger efficiency than micro and small non-exporters

Table 7.4 Bias-corrected efficiency differences between exporters and non-exporters

	Average bias-corrected efficiency		Number of observations		H_0: Exporters and non-exporters have equal efficiency distributions		H_0: Exporters have a higher efficiency than non-exporters	
Year	Exporters	Non-exporters	Exporters	Non-exporters	Statistic	*P*-value	Statistic	*P*-value
2000	0.800 (0.084)	0.728 (0.122)	65	199	0.3110	0.000	−0.0101	0.990
2001	0.713 (0.144)	0.618 (0.140)	66	253	0.3472	0.000	−0.0250	0.937
2002	0.687 (0.113)	0.617 (0.139)	75	287	0.2398	0.002	−0.0209	0.949
2003	0.767 (0.097)	0.719 (0.100)	78	289	0.2899	0.000	−0.0094	0.989
2004	0.712 (0.164)	0.654 (0.137)	79	293	0.3331	0.000	−0.0486	0.745
2005	0.716 (0.115)	0.679 (0.113)	80	292	0.1940	0.018	−0.0204	0.949
2006	0.713 (0.130)	0.646 (0.136)	80	292	0.2702	0.000	−0.0068	0.994
2007	0.693 (0.128)	0.648 (0.140)	77	268	0.2032	0.014	−0.0187	0.959
2008	0.702 (0.116)	0.670 (0.123)	90	290	0.1897	0.014	−0.0640	0.570
2009	0.629 (0.146)	0.545 (0.135)	80	276	0.3236	0.000	−0.0065	0.995

The values presented in brackets are standard deviations

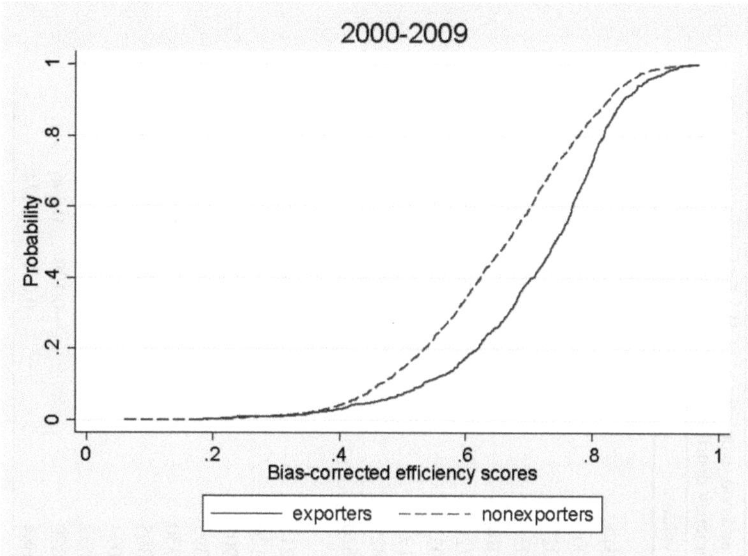

Fig. 7.1 Cumulative distributions of bias-corrected efficiency scores: exporters versus non-exporters

cannot be rejected at any reasonable significance level, even in the years when first hypothesis cannot be rejected. Therefore, although the differences in efficiency between exporters and non-exporters in these years are rather modest, they still favour exporters. On the other hand, for the groups of medium and large firms, the equality of efficiency distributions of exporters versus non-exporters was not rejected at any reasonable significance level, except for the year 2006. However, the small efficiency differences between exporters and non-exporters still favour medium and large exporters as shown by the test results of the second hypothesis.

7.5 Conclusions

This chapter measured the technical efficiency of Spanish dairy processing firms and statistically tested for differences in technical efficiency between firms of different size and between firms that operate in international markets versus those that are not. The empirical application focuses on a sample of Spanish dairy processing firms over the period 2001–2009.

The results showed that technical efficiency increases with the size of dairy processing firms, that is, larger dairy processing firms were, on average, more efficient than smaller dairy processing firms. Furthermore, within the group of micro and small processing firms, those firms that were exporting were more efficient than firms that do not export. The distribution of technical efficiency of

Table 7.5 Bias-corrected efficiency differences between exporters and non-exporters within size classes

	Average bias-corrected efficiency		Number of observations		H_0: Exporters and non-exporters have equal efficiency distributions		H_0: Exporters have a higher efficiency than non-exporters	
Year	Exporters	Non-exporters	Exporters	Non-exporters	Statistic	*P*-Value	Statistic	*P*-Value
Micro and small exporting firms versus micro and small non-exporting firms								
2000	0.781 (0.093)	0.719 (0.124)	43	177	0.238	0.040	−0.006	0.998
2001	0.690 (0.163)	0.612 (0.141)	44	238	0.316	0.001	−0.053	0.811
2002	0.669 (0.122)	0.609 (0.139)	51	268	0.225	0.027	−0.022	0.958
2003	0.751 (0.106)	0.714 (0.100)	53	269	0.226	0.021	−0.015	0.980
2004	0.674 (0.178)	0.645 (0.136)	56	271	0.232	0.014	−0.078	0.572
2005	0.679 (0.112)	0.675 (0.112)	51	271	0.105	0.728	−0.065	0.699
2006	0.673 (0.135)	0.641 (0.137)	52	271	0.215	0.035	−0.019	0.971
2007	0.670 (0.143)	0.644 (0.141)	42	249	0.184	0.177	−0.028	0.945
2008	0.665 (0.128)	0.665 (0.122)	44	267	0.102	0.829	−0.077	0.637
2009	0.551 (0.141)	0.534 (0.131)	43	254	0.124	0.620	−0.079	0.634
Medium and large exporting firms versus medium and large non-exporting firms								
2000	0.835 (0.048)	0.804 (0.066)	22	22	0.318	0.215	−0.046	0.956
2001	0.760 (0.078)	0.720 (0.075)	22	15	0.394	0.126	−0.024	0.990
2002	0.727 (0.082)	0.727 (0.073)	24	19	0.173	0.908	−0.103	0.798
2003	0.800 (0.066)	0.785 (0.077)	25	20	0.270	0.393	−0.110	0.764
2004	0.803 (0.068)	0.771 (0.088)	23	22	0.283	0.330	−0.059	0.924
2005	0.780 (0.090)	0.735 (0.108)	29	21	0.278	0.305	−0.031	0.977
2006	0.787 (0.077)	0.720 (0.107)	28	21	0.357	0.094	0.000	1.000
2007	0.721 (0.103)	0.696 (0.115)	35	19	0.212	0.637	−0.051	0.938
2008	0.737 (0.091)	0.725 (0.128)	46	23	0.152	0.870	−0.109	0.696
2009	0.719 (0.087)	0.673 (0.118)	37	22	0.193	0.684	−0.031	0.974

The values presented in brackets are standard deviations

the small- and micro-sized exporters stochastically dominates the distribution of the non-exporters in most years. For medium-sized and large firms, the technical efficiency of exporters was significantly higher than of non-exporters, but the distribution of technical efficiency of exporters does not stochastically dominate the distribution of non-exporters.

The evidence in this chapter suggests that size has a positive impact on the performance of Spanish dairy processing firms. Also, participation in export markets matters for the performance; this holds particularly for the small and micro firms. These results imply that larger firms and firms that operate in international markets are in a better position to cope with future challenges following from globalization, market liberalization and changing consumer preferences.

References

Aw BY, Hwang A (1995) Productivity and the export market: a firm-level analysis. J Dev Econ 47:313–332

Aw BY, Chen X, Roberts MJ (1997) Firm level evidence on productivity differentials, turnover, and exports in Taiwanese manufacturing. NBER Working Paper 6235

Aw BY, Chung S, Roberts MJ (2000) Productivity and turnover in the export market: micro evidence from Taiwan and South Korea. World Bank Econ Rev 14:65–90

Banker RD, Charnes A, Cooper WW (1984) Some models for estimating technical and scale inefficiencies in data envelopment analysis. Manag Sci 30:1078–1092

Bernard AB, Jensen JB (1995) Exporters, jobs and wages in U.S. manufacturing, 1976–1987, the brooking papers on economic activity. Microeconomics 1995:67–112

Boel MF (2008) The cap health check: good news for variety in European farming. Speech at the Committee of the Regions meeting, Brussels, Belgium. Retrieved December 10, 2011, from http://brussels.ifa.ie/ifa/brussels/media/permanent%20info/CAP%20Health%20Check%20good%20news%20for%20variety%20in%20European%20Farming.pdf

Boyle GE (2004) The economic efficiency of Irish dairy marketing cooperatives. Agribusiness 20:143–153

Charnes A, Cooper WW, Rhodes E (1978) Measuring the efficiency of decision making units. Eur J Oper Res 2:429–444

Chen T, Tang D (1987) Comparing technical efficiency between import-substitution-oriented and export-oriented foreign firms in a developing economy. J Dev Econ 26:277–289

Clerides SK, Lach S, Tybout JR (1998) Is learning-by-exporting important? Micro-dynamic evidence from Colombia, Mexico and Morocco. Q J Econ CXIII:903–947

Coelli T, Rao DSP, Battese GE (2005) An introduction to efficiency and productivity analysis. Springer, New York

Conover WJ (1971) Practical nonparametric statistics. Wiley, New York

Delgado M, Fariñas JC, Ruano S (2002) Firm productivity and export markets: a non-parametric approach. J Int Econ 57:397–422

Devlieghere F, Vermeiren L, Debevere J (2004) New preservation technologies: possibilities and limitations. Int Dairy J 14:273–285

Doucouliagos H, Hone P (2001) The efficiency of the Australian dairy processing industry. Aust J Agric Resour Econ 44:423–438

Eurostat (2012) http://epp.eurostat.ec.europa.eu/portal/page/portal/eurostat/home

Jensen JB, Wagner J (1997) Exports and success in German manufacturing. Weltwirtschaftliches Archiv 133:134–157

Kleibeuker J (2007) Dairy production faces unique challenges, expert claims. In: Merrett N (ed) Breaking news on dairy processing & markets. Retrieved September 20, 2007, from http://www.dairyreporter.com/Industry-markets/Dairy-production-faces-unique-challenges-expert-claims

Leibenstein H (1976) Beyond economic Man. Harvard University Press, Cambridge, MA

Majumdar S (1997) The impact of size and age on firm-level performance: some evidence from India. Rev Ind Organ 12:231–241

Martin-Marcos A, Suarez-Galvez C (2000) Technical efficiency of Spanish manufacturing firms: a panel data approach. Appl Econ 32:1249–1258

Simar L, Wilson PW (1998) Sensitivity analysis of efficiency scores: how to bootstrap in nonparametric frontier models. Manag Sci 44:49–61

Simar L, Wilson PW (2000) A general methodology for bootstrapping in non-parametric frontier models. J Appl Stat 27:779–802

Singh S, Coelli T, Fleming E (2001) Performance of dairy plants in the cooperative and private sectors in India. Ann Public Coop Econ 72:453–479

Soboh RAME, Oude Lansink A, Giesen G, Van Dijk G (2009) Performance measurement of the agricultural marketing cooperative: the gap between theory and practice. Rev Agric Econ 31:446–469

Soboh R, Oude Lansink A, Van Dijk G (2012) Efficiency of investor owned firms and cooperatives revisited. J Agric Econ 63:142–157

Spanish Statistical Office (2012) (Instituto Nacional de Estadística) www.ine.es

Sueyoshi T, Aoki S (2001) A use of a nonparametric statistic for DEA frontier shift: the Kruskal and Wallis rank test. Omega-Int J Manag Sci 29:1–18

Wilson PW (2008) FEAR: a software package for frontier efficiency analysis with R. Socio-Econ Plan Sci 42:247–254

Chapter 8
Inefficiency in Animal Production: A Parametric Approach

Emiliana Silva and Fátima Venâncio

Abstract A stochastic frontier approach (SFA) was estimated for three types of farms using Frontier Software. The groups of farms were created using cluster analysis and SPSS Statistical package for the Social Sciences. The Frontier Program allowed the estimation of efficiency (model I) and inefficiency models (model II). The efficiency in Faial (Azores) island farms was higher than 82%, and the most important inefficiency variables were subsidies, equipment amortization and small dimension.

Keywords Efficiency • Stochastic frontier production • Azores • Cluster

8.1 Introduction

The objectives of this chapter are (1) to estimate the technical efficiency (by stochastic frontier production) of the animal production farms of Faial (Azores, Portugal) and (2) to find the variables that influence the technical inefficiency.

In a first stage similar farms were obtained (clusters) since it was suspected that heterogeneity could affect the efficiency; in the next stage, a stochastic frontier production (efficiency and inefficiency models) was estimated for each cluster.

The output of the efficiency model was the income and the inputs were the area, the main variables and the fixed expenses. The variables considered as causing inefficiency were as follows: animal number (AN), subvention, beef sales, amortization, farm ownership type, economic dimension units (EDU) and the experience of the farmer (number of working years in the farm).

E. Silva (✉) • F. Venâncio
CEEAplA, Departamento de Ciências Agrárias, Universidade dos Açores, Rua Capitão João de Ávila, 9700-042 Angra do Heroísmo, Açores, Portugal
e-mail: emiliana@uac.pt; fvenancio@sapo.pt

A.B. Mendes et al. (eds.), *Efficiency Measures in the Agricultural Sector: With Applications*, DOI 10.1007/978-94-007-5739-4_8,

The Frontier version 4.1, developed by Coelli (1996), was selected to estimate the models. The stochastic frontier approach (SFA) is a parametric approach which was originally and independently proposed by Aigner et al. (1977) and Meeusen and van den Broeck (1977) as referred by Battese and Coelli (1988). The SFA considers that an error term has two components: one to account for random effects and another to account for the technical inefficiency effects (Coelli 1995).

The efficiency model (model I) was proposed by Battese and Coelli (1992), and the u_{it} is the technical inefficiency variable which is defined by

$$u_{it} = \{\exp[-\eta(t - T)]\}\, u_i, \quad \text{with } i = 1, \ldots, N, \quad \text{and } t = 1, \ldots, T, \tag{8.1}$$

where η is an unknown parameter to be estimated and μ_i with $i = 1, 2, \ldots, N$, are independent and identically distributed non-negative random variables, which are assumed to account for technical inefficiency in production and obtained by truncations (at zero) of the normal distribution with unknown mean and unknown variance, σ^2.

This model specifies that the technical inefficiency effects for the sample in the earlier periods of the panel are a deterministic exponential function of the inefficiency effects for the corresponding firms in the last years of the panel (i.e., $u_{it} = u_i$, given the data for the ith firm is available in the period, T).

Given Eq. (8.1), the expectation of the mean technical efficiency is

$$\mathrm{TE} = \exp(-u_i) \tag{8.2}$$

and it is estimated considering the technical inefficiency effects (u_i).

Model I only permits to determine the technical efficiency in order to learn more about inefficiency. Battese and Coelli (1995) presented model II (inefficiency model) which permits to incorporate in model I the variable that could cause inefficiency in the firms. In this model the technical inefficiency effects are defined by

$$u_{it} = z_{it}\delta + w_{it}, \quad \text{and } i = 1, \ldots, N, \quad \text{and } t = 1, \ldots, T, \tag{8.3}$$

where z_{it} is $(1 \times M)$ vector of explanatory variables associated with technical inefficiency effects,

δ is $(M \times 1)$ vector of unknown parameters to be estimated and w_{it} are the unobservable random errors which are assumed to be independently distributed and obtained by truncation of normal distribution with unknown mean and variance, σ^2, such that u_{it} is non-negative ($w_{it} \geq -z_{it}\delta$).

Some examples of surveys using these models are Tauer and Mishra (2006), Hasnah Fleming and Coelli (2004), Lawson et al. (2004), Pascoe et al. (2001), Torres et al. (2002), Munzir and Heidhues (2002), Reinhard et al. (1999, 2000), Alvarez and Gonzalez (1999), Webster et al. (1998), Franco (2001) and Daryanto et al. (2002). In agriculture survey, we have Lawson et al. (2004), Venâncio and Silva (2004), Battese and Coelli (1988), Battese and Sumiter (1997), Brummer (2001) and Hallam and Machado (1996).

8.2 Materials and Methods

The FADN (Farm Agriculture Data Network) was used for Azores database (47 units) from 1996 to 1999. The data was analysed using current prices. The variables used in the cluster analysis were (1) income (subventions, milk and beef sales), (2) dimension and (3) variable expenses (rent equipments, conservation and reparations of equipment; fuel and oil; concentrate food and other specific expenses; fertilizers and general expenses).

The main characteristic of the farm cluster was as follows: (1) cluster A (smaller dimension, smaller number of animals, smaller incomes, subvention and beef sales' enormous importance), (2) cluster B (medium and larger dimension, smaller number of animals, mainly dairy production, medium income, smaller importance of earning subvention and beef sales, larger expenses of concentrate animals, higher investment) and (3) cluster C (medium and larger dimension, mainly dairy production, larger animal number, larger expenses in concentrate and fertilizers, larger incomes).

The efficiency and inefficiency models were applied to each cluster. In the efficiency model, it was necessary to group some expense variable to adequate the variable and observation number and to estimate SFA in translogarithmic and Cobb-Douglas functional forms. The inputs of production function are represented by four variables: (1) dimension (SAU), hectares; (2) expenses with animals per hectare (AN/HA); (3) expenses with equipment per hectare (EQ/HA); and (4) other expenses per hectare (OC/HA).

The variables' selection for the inefficiency models was based on previous works of Hallam and Machado (1996) and Puig-Junoy and Argilés (2000). These eight variables were as follows: (1) TIME *Trend*, (2) ANHA (animals per hectare), (3) SUBS (percent of subvention in income), (4) AS (percent of beef sales in income), (5) AM (amortization), (6) OW (ownership farm type *Dummy*) (OW = 1, rent, or OW = 0, not rent), (7) UDE (economic dimension unit) and (8) AG (experience age).

The next Figs. 8.1 and 8.2 present the steps followed in the methodology and the sequence of estimating the efficiency and inefficiency models.

The main procedures were (1) to define the specifications of each model using a statistic LR test, likelihood ratio tests; (2) to present the efficiency measures per cluster; (3) to define the final equations for both models (I and II) per cluster; and (4) to analyse the variables that caused inefficiency (Fig. 8.1).

Figure 8.2 shows the Frontier Software (Coelli 1996) step's procedure: (1) ordinary least squares (OLS) estimated the obtained function; all β estimators except the intercept were unbiased; (2) a two-phase grid search of γ was conducted, with the β parameters (except β_0) set to the OLS values and the β_0 and σ^2 parameters adjusted according to the corrected ordinary least squares formula presented in Coelli (1995). All other parameters (μ, α and δ's) were set to zero in this grid search; and (3) the values selected in the grid search were used as starting values in an iterative procedure to obtain the final maximum-likelihood estimates.

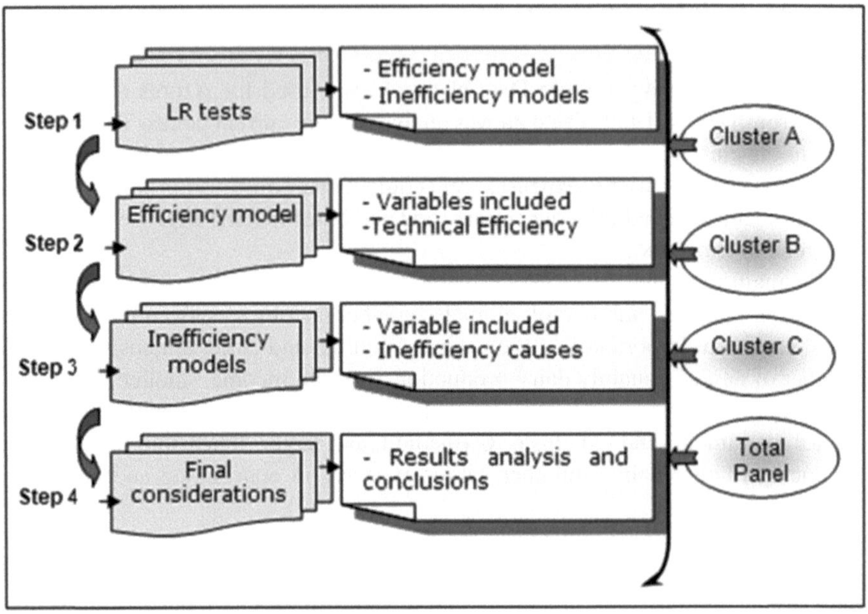

Fig. 8.1 The steps in the model applications

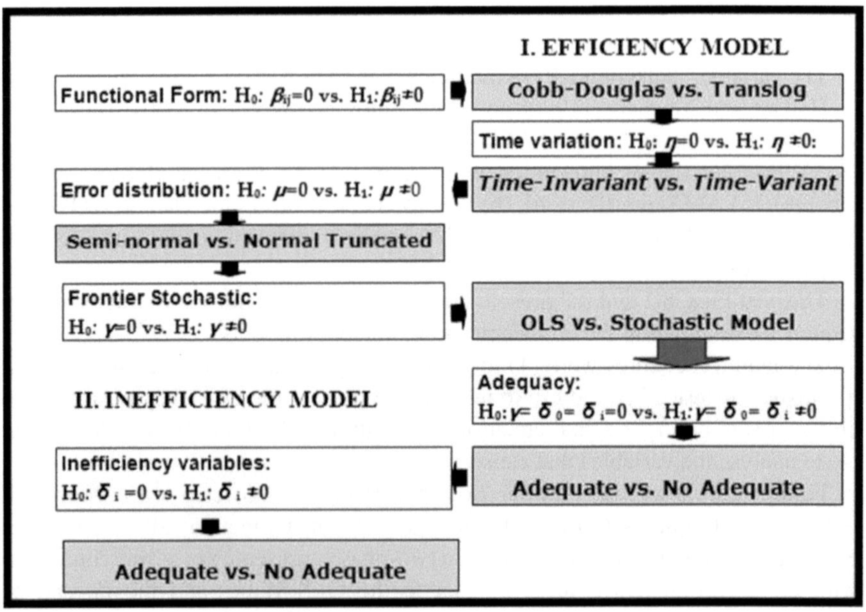

Fig. 8.2 LR tests and models I and II

The grid search value results were used in an interactive procedure to obtain LR test estimates. Critical chi-squared values were obtained by Kodde and Palm (1986) and used to test if the estimated γ and the variables explaining inefficiency jointly equalled 0.

The statistical tests were made according to the following: (a) the functional form that best adjust to the data, (b) the choice of the time-variant or time-invariant model, (c) the distribution of error model, (d) the adequate model to stochastic production according to the used data, (e) to adequate the inefficiency model to the data and (f) to adequate the inefficiency variable.

8.3 Results and Discussion

The following Tables 8.1, 8.2 and 8.3 and Figs. 8.3 and 8.4 show the main results for models (I and II) per cluster and by total panel.

Table 8.1 summarizes the results. In the efficiency model, the functional form is the translog, in cluster A and B, and Cobb-Douglas in cluster C and total panel. The time variation is variant only for the total panel. The error distribution is normal truncated to the total panel and semi-normal for all clusters. For all cases, there is a stochastic frontier. In the efficiency model, all cases have an adequate model, but in cluster A, the variables of inefficiency are not adequate.

Table 8.2 shows that the σ^2 parameter (variance of inefficiency term, u, and the residual term, ν) is significant for all the analysed groups. The defined γ has a value between 0 and 1: if there is inefficiency; if it is different from 0, then the error term expresses the traditional random variation that is uncontrolled by the farmer. There is variance between all different parameters of the model efficiency. The parameter γ (ratio of total model variance and the variance relative to the inefficiency firms) has lower values than cluster B but statistically significant. As a complementary analysis of this hypothesis test, the value γ was estimated to observe the existence or not of the stochastic frontier.

The γ values for clusters A, B and C were 0.766, 0.479 and 0.892, accordingly (Table 8.2). This value for cluster B though were relatively low (*between* 0 and 1) shows technical inefficiency in production processes and statistically significant at 5 and 1%, accordingly. Although this results revealed that only 47.9% of inefficiency of these farms were due to factors inherent of farms and the remaining were due to random and external factors of farms and not directly controlled, this result must be carefully considered because γ is relatively low.

The higher values of γ for clusters A and C confirm the results obtained by LR test, about the adequacy of model I to our survey data. As a consequence, the average production function (OLS) is less adequate to the frontier production

The η and μ values resulted from the specifications of statistic tests, and when they have zero values, the model adopted is *time-invariant* with semi-normal distribution; when they have a value different from zero, the model adopted is *time-variant* with error tem distribution truncated.

Table 8.1 Tests of hypotheses for coefficients of the technical efficiency variables in the stochastic frontier production function by cluster and total panel

		λ	Critical value (5%)	Decision
C	Model I			
l	H_0: $\beta_{ij} = 0, i \leq j = 1, 2, \ldots, 10$	71.38	18.31	Rejected H_0
u	H_0: $\eta = 0$	3.32	3.84	Accepted H_0
s	H_0: $\mu = 0$	0	3.84	Accepted H_0
t	H_0: $\gamma = 0$	13.82	7.05	Rejected H_0
e	Model II			
r	H_0: $\gamma = \delta_0 = \delta_i = 0, i = 1, \ldots, 8$	2, 271.58	16.27	Rejected H_0
A	H_0: $\delta_1 = \cdots = \delta_8 = 0$	13.76	15.51	Accepted H_0
C	Model I			
l	H_0: $\beta_{ij} = 0, i \leq j = 1, 2, \ldots, 10$	60.14	18.31	Rejected H_0
u	H_0: $\eta = 0$	−1.78	3.84	Accepted H_0
s	H_0: $\mu = 0$	−0.58	3.84	Accepted H_0
t	H_0: $\gamma = 0$	22.16	7.05	Rejected H_0
e	Model II			
r	H_0: $\gamma = \delta_0 = \delta_i = 0, i = 1, \ldots, 8$	2, 039.80	16.27	Rejected H_0
B	H_0: $\delta_1 = \cdots = \delta_8 = 0$	24.1	15.51	Rejected H_0
C	Model I			
l	H_0: $\beta_i = 0, i = 1, 2, 3, 4$	16.66	18.31	Accepted H_0
u	H_0: $\eta = 0$	0.24	3.84	Accepted H_0
s	H_0: $\mu = 0$	2.24	3.84	Accepted H_0
t	H_0: $\gamma = 0$	44.88	7.05	Rejected H_0
e	Model II			
r	H_0: $\gamma = \delta_0 = \delta_i = 0, i = 1, \ldots, 8$	996.82	16.27	Rejected H_0
C	H_0: $\delta_1 = \cdots = \delta_8 = 0$	15.72	15.51	Rejected H_0
T	Model I			
o	H_0: $\beta_i = 0, i = 1, 2, 3, 4$	1.4	18.31	Accepted H_0
t	H_0: $\eta = 0$	5.08	3.84	Rejected H_0
a	H_0: $\mu = 0$	7.96	3.84	Rejected H_0
l	H_0: $\gamma = 0$	15.72	7.05	Rejected H_0
	Model II			
	H_0: $\gamma = \delta_0 = \delta_i = 0, i = 1, \ldots, 8$	6, 100.46	16.27	Rejected H_0
	H_0: $\delta_1 = \cdots = \delta_8 = 0$	35.94	15.51	Rejected H_0

Table 8.2 Estimative of some parameters of SFA – model I

Parameter	Cluster A	Cluster B	Cluster C	Total panel
$\sigma^2 = \sigma_\mu^2 + \sigma_\upsilon^2$	0.091***	0.016***	0.043*	0.066***
$\gamma = \sigma_\mu^2/\sigma^2$	0.766***	0.479**	0.892***	0.391***
μ	0	0	0	0.321***
η	0	0	0	0.076***

***($P < 0.01$)*; ** ($P < 0.05$);* P($P < 0.10$)*

Table 8.3 Maximum-likelihood estimates for parameters of the stochastic frontiers for the clusters and total panel

Variable	Parameter	Cluster A	Cluster B	Cluster C	Total panel
Constant	δ_0	−0.179	1.454	1.606**	1.271
Time Trend	δ_1	−0.031	−0.017	0.018	−0.055***
ln (AN/HA)	δ_2	−0.185	0.003	0.166	−0.241**
ln (SUBS)	δ_3	0.001	0.145***	0.204	0.045***
ln (AS)	δ_4	−0.109	−0.029	−0.147*	−0.058***
ln (AM)	δ_5	0.083	−0.051	0.205**	0.105***
OW Dummy	δ_6	−0.453***	−0.025	−0.082	−0.117**
ln (UDE)	δ_7	−0.009	−0.146	−0.688*	−0.201***
ln (AG)	δ_8	0.193	−0.106	−0.012	−0.196
	$\sigma^2 = \sigma_\mu^2 + \sigma_v^2$	0.032***	0.008***	0.014***	0.048***
	$\gamma = \sigma_\mu^2/\sigma^2$	0.216	0.999***	0.999***	0.038

***(*P* < 0.01)*; ** (*P* < 0.05);* *P*(*P* < 0.10)*

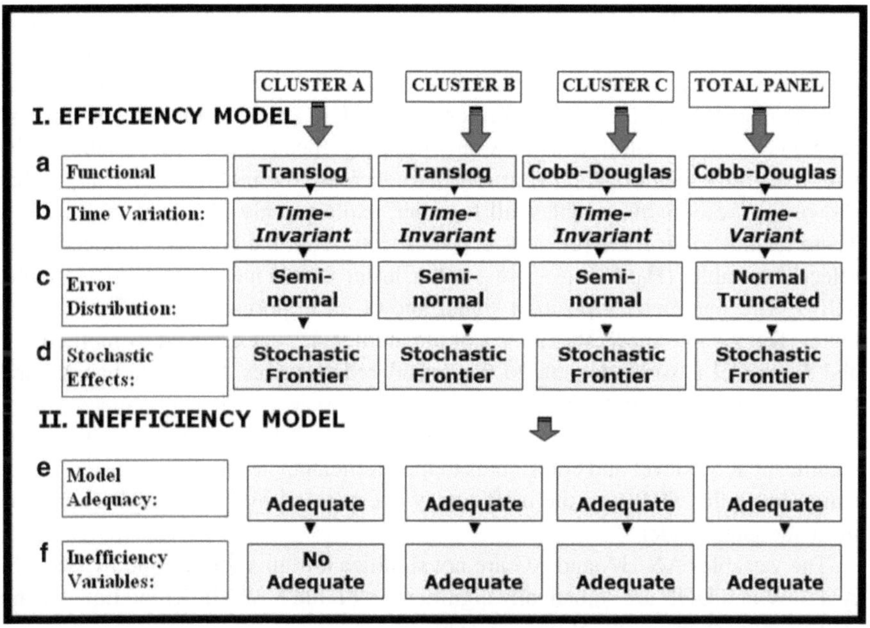

Fig. 8.3 Result models using the efficiency and inefficiency models

The three clusters for model I have similar values of efficiency intervals. They have more farms in the interval of 0.80–0.89 and superior to 0.90. Cluster A showed the efficiency lower values and that there are more farms in the groups inferior to 0.80. The difference between clusters B and C is at a lower level of efficiency for cluster C, which restrained the difference of the final values of both clusters.

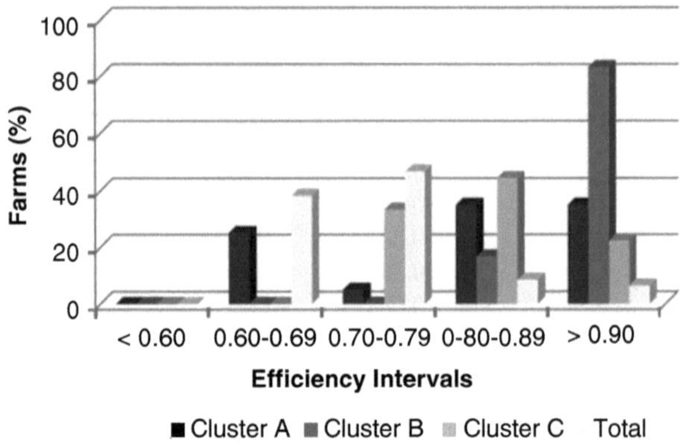

Fig. 8.4 Distribution of the technical efficiency clusters by using the efficiency interval

The average efficiencies for clusters A, B and C are 82.0, 93.2 and 85.1%, accordingly, and the most relevant efficiency intervals are superior to 80% (Fig. 8.4).

The results of inefficiency model – Model II – are shown in Table 8.3. The statistical variables are subvention, beef sales, amortization, ownership and economic dimension unit. Cluster A only has a significant inefficient variable (OW) with a negative signal, which contributes to decrease the inefficiency. The γ value shown for these eight variables, all together, explains only 21.6% of inefficiency. These results confirm the previous statistics tests about the adequate models to the selected variables (H_0: $\delta_1 = \cdots = \delta_8 = 0$). Cluster B only has a significant variable, SUBS. The parameter's positive signal shows its importance as an inefficiency factor. The γ value shows that 99.9% of the inefficiency is explained by inefficiency and the model is well adequate to the considered variables. Cluster C has AS and UDE, which were significant at a 10% significance level. Their parameter values have a negative signal and they contribute to decrease the inefficiency. The AM is significant at 5% level and contributes to the inefficiency level of this cluster. The γ value shows that 99.9% of the inefficiency is explained by these variables and the most adequate model.

The variables AN/HA and AG are not significant in this survey, and SUBS, AS, AM, OW and UDE are statistically significant y (Table 8.3). The Trend time has no significance (short period of time, 4 years).

8.4 Conclusions

The SFA model was more adequate than the traditional models of average production (OLS). This fact was demonstrated by the great adequacy of the models using the translog production function or Cobb-Douglas function.

To get a frontier in both models, with a different production function (translog and Cobb-Douglas) for each group of analysis, the input and output were the same for the three farm types; we verified that the stochastic frontier is sensible to the used data. When the variables are grouped in different ways, they result in a different frontier production and level of efficiency.

The levels of efficiency are constant and similar throughout time, and their values are 82, 93.2 and 85.1% accordingly for the clusters A, B and C.

The variables that cause inefficiency are subvention and equipment amortization. The variables with a major level of connection to efficient farms are lower rent fields, dimension and beef sales. The biggest farms seem more efficient, such as observed in Hallam and Machado (1996).

References

Aigner D, Lovell C, Scmidt P (1977) Formulation and estimation of stochastic frontier production function models. J Econ 6(1):21–37

Alvarez A, Gonzalez E (1999) Using cross-section data to adjust technical efficiency indexes estimated with panel data. Am J Econ 81:894–901

Battese G, Coelli T (1988) Prediction of firm-level technical efficiencies with a generalised frontier production function and panel data. J Econ 38:387–399

Battese G, Coelli T (1992) Frontier production functions. Technical efficiency and panel data: with application to paddy farmers in India. J Product Anal 3:153–169

Battese G, Coelli T (1995) A model for technical inefficiency effects in a stochastic frontier production function for panel data. Empir Econ 20:325–332

Battese G, Sumiter B (1997) Functional forms of stochastic frontier production function and models for technical inefficiency effects: a comparative study for wheat farmers in Pakistan. J Product Anal 8:395–414

Brummer B (2001) Estimating confidence intervals for technical efficiency: the case of private farms in Slovenia. Eur Rev Agric Econ 28(3):285–306

Coelli T (1995) Recent developments in frontier modelling and efficiency measurement. Aust J Agric Econ 39:219–245

Coelli T (1996) A guide to FRONTIER Version 4.1: a computer program for stochastic frontier production and cost function estimation, CEPA Working paper no. 96/07. Department of Econometrics, University of New England, Armidale

Daryanto H, Battese G, Fleming E (2002) Technical efficiencies of rice farmers under different irrigation systems and cropping seasons in West Java, University of New England, Asia Conference on Efficiency and Productivity Growth. Retrieved May 20, 2003 from http://www.sinica.edu.tw/~teps/A1-1.pdf

Franco F (2001) Eficiência Comparada Dos Serviços Hospitalares: O Método de Fronteira Estocástica, Tese de Mestrado, Universidade dos Açores

Hallam D, Machado F (1996) Efficiency analysis with panel data – a study of Portuguese dairy farms. Eur Rev Agric Resour Econ 23(1):79–93

Hasnah Fleming E, Coelli T (2004) Assessing the performance of a nucleus estate and smallholder scheme for oil palm production in West Sumatra: a stochastic frontier analysis. Agric Syst 79(1):17–30

Kodde D, Palm F (1986) Wald criteria for jointly testing equality and inequality restrictions. Econometrica 54(5):1243–1248

Lawson L, Bruun J, Coelli T, Agger J, Lund M (2004) Relationships of efficiency to reproductive disorders in Danish milk production: a stochastic frontier analysis. J Dairy Sci 87:212–224

Meeusen W, van der Broeck J (1977) Efficiency estimation from Cobb-Douglas production functions with composed error. Int Econ Rev 18:435–444

Munzir A, Heidhues F (2002) Towards a technically efficient production in rural aquaculture, case study at Lake Maninjau. Indonesia, International Symposium Sustaining Food Security and Managing Natural Resources in Southeast Asia – Challenges for the 21st Century, Chiang Mai, Tailândia. Retrieved July 21, 2003 from http://www.uni-hohenheim.de/symposium2002/pa_full/Full-Pap-S3B-3_Munzir.pdf

Pascoe S, Hassaszahed P, Anderson J, Korsbrekke K (2001) Economic versus physical input measures in the analysis of technical efficiency in fisheries, XII Conference of the European Association of Fisheries Economists, Italia. http://www.eafe-fish.org/conferences/salerno/papers/paper04_seanpaschoe .doc. 2003-04-25

Puig-Junoy J, Argilés J (2000) Measuring and explaining farm inefficiency in a panel data set of mixed farms, Universidade Pompeu Fabra, Espanha. Retrieved March 21, 2003 from http://www.econ.upf.es/deehome/what/wpapers/postscripts/503.pdf

Reinhard S, Lovell K, Thijssen G (1999) Econometric estimation of technical and environmental efficiency: an application to Dutch dairy farm. Am J Agric Econ 81:44–60

Reinhard S, Lovell K, Thijssen G (2000) Environmental efficiency with multiple environmentally detrimental variable: estimated SFA e DEA. Eur J Oper Res 121:287–303

Tauer L, Mishra A (2006) Dairy farm cost efficiency. J Dairy Sci 89:4937–4943

Torres J, Basch M, Vergara S (2002) Eficiência Técnica Y Escalas de Operación en Pesca Pelágica: Un Análisis de fronteras estocásticas, Universidade do Chile. Retrieved April 21, 2003 from http://www.ilades.cl/economia/Publicaciones/ser_inv/inv137.pdf

Venâncio F, Silva E (2004) A Eficiência de Explorações Agro-pecuárias dos Açores: uma abordagem paramétrica, Actas XIV Jornadas Luso – Espanholas de Gestão Científica

Webster R, Kennedy S, Johnson L (1998) Comparing techniques for measuring the efficiency and productivity of Australian private hospitals, Working papers (98/3) in econometrics and applied statistics, Australian Bureau of Statistics. Retrieved May 09, 2003 from http://www.abs.gov.au/websitedbs/D3110122.NSF/0/31f3a2cdb3dbbf85ca25671b001f1910/$FILE/13510_Nov98.pdf

Chapter 9
Azorean Agriculture Efficiency by PAR

Armando B. Mendes, Veska Noncheva, and Emiliana Silva

Abstract The producers always aspire at increasing the efficiency of their production process. However, they do not always succeed in optimising their production. In the last years, the interest on Data Envelopment Analysis (DEA) as a powerful tool for measuring efficiency has increased. This is due to the large amount of data sets collected to better understand the phenomena under study and, at the same time, to the need of timely and inexpensive information.

The "Productivity Analysis with R" (PAR) framework establishes a user-friendly data envelopment analysis environment with special emphasis on variable selection, aggregation, summarisation and interpretation of the results. The starting point is the following R packages: DEA (Diaz-Martinez and Fernandez-Menendez 2008) and FEAR (Wilson 2008). The DEA package performs some models of data envelopment analysis presented in Cooper et al. (2007). FEAR is a software package for computing nonparametric efficiency estimates and testing hypotheses in frontier models. FEAR implements the bootstrap methods described in Simar and Wilson (2000).

PAR is a software framework using a portfolio of models for efficiency estimation and also providing results explanation functionality. PAR framework has been developed to distinguish between efficient and inefficient observations and

A.B. Mendes (✉)
CEEAplA, Departamento de, Matemática, Universidade dos Açores, Rua da Mãe de Deus, 9500-801 Ponta Delgada, Açores, Portugal
e-mail: amendes@uac.pt

V. Noncheva
CEEAplA, Faculty of Mathematics and Informatics, University of Plovdiv, "PaisiiHilendarski" Plovdiv, Bulgaria

E. Silva
CEEAplA, Departamento de Ciências Agrárias, Universidade dos Açores, Rua Capitão João de Ávila, 9700-042 Angra do Heroísmo, Açores, Portugal
e-mail: emiliana@uac.pt

A.B. Mendes et al. (eds.), *Efficiency Measures in the Agricultural Sector: With Applications*, DOI 10.1007/978-94-007-5739-4_9,

to explicitly advise the producers about possibilities for production optimisation. PAR framework offers several R functions for a reasonable interpretation of the data analysis results and text presentation of the obtained information. The output of an efficiency study with PAR software is self-explanatory.

We are applying PAR framework to estimate the efficiency of the agricultural system in Azores (Mendes et al. 2009). All Azorean farms will be clustered into homogeneous groups according to their efficiency measurements to define clusters of "good" practices and cluster of "less good" practices. This makes PAR appropriate to support public policies in agriculture sector in Azores.

Keywords Productivity Analysis with R • Data Envelopment Analysis • Efficiency of Azorean farms

9.1 Introduction

DEA makes it possible to identify efficient and inefficient units in a framework where results are considered in their particular context. The units to be assessed should be relatively homogeneous and were originally called Decision-Making Units (DMUs). DEA is an extreme point method and compares each DMU with only the "best" DMUs.

DEA can be a powerful tool when used wisely. A few of the characteristics that make it powerful are:

- DEA can handle multiple input and multiple output models.
- DMUs are directly compared against a peer or combination of peers.
- Inputs and outputs can have very different units. For example, one variable could be in units of lives saved and another could be in units of dollars without requiring an a priori trade-off between the two.

The same characteristics that make DEA a powerful tool can also create problems. An analyst should keep these limitations in mind when choosing whether or not to use DEA:

- Since DEA is an extreme point technique, noise such as measurement error can cause significant problems.
- DEA is good at estimating "relative" efficiency of a DMU, but it converges very slowly to "absolute" efficiency. In other words, it can tell you how well you are doing compared to your peers but not compared to a "theoretical maximum".

PAR combines DEA with different statistical methods. DEA is applied to distinguish between efficient and inefficient observations of performances. Different statistical methods are applied to assist DEA. For example, canonical correlation analysis assists DEA with both variable aggregation and variable selection. PAR methodology is implemented in R. The output of the PAR computer program is self-explanatory.

At first, we will define the performance of a farm. A natural measure of performance is a productivity ratio: the ratio of outputs to inputs, where larger values of this ratio are associated with better performance. Performance is a relative concept. For example, the performance of the meat farm in 2008 could be measured relative to its 2007 performance or it could be measured relative to the performance of another farm in 2008. This farm can also analyse the relative performance of units within the farm.

9.2 PAR: A Tool for Measuring Efficiency of Azorean Farms

9.2.1 Basic Term Definitions

We are going to provide some informal definitions of the following terms.

9.2.1.1 Productivity

Productivity can be simply defined as the ratio between outputs and inputs of an economic system. When we refer to productivity, we are referring to total farm productivity, which is a productivity measure involving all factors of production (all inputs and all outputs). The land productivity yields in farming are a partial measure of productivity. The partial productivity measures can provide a misleading indication of overall productivity when considered in isolation.

9.2.1.2 Production Frontier Line

The production frontier line may be used to define the relationship between the input and output. The production frontier represents the maximum output attainable from each input level. It reflects the current state of technology in the farm. Farms operate either on that frontier, if they are technically efficient, or beneath the frontier, if they are technically inefficient.

Efficiency frontier represents a standard of performance that the firms not on the frontier could try to achieve. Firms on the frontier are 100% efficient.

Note that this does not mean that the performance of the DMUs on the efficiency frontier cannot be improved. It may or may not be possible. However, the available data does not give any idea on the extent to which their performance can be improved.

The DMUs on the efficiency frontier are the best DMUs with the data that we have. As we do not have another DMU having better performance, we should assume that these are the best achievable performances. We rate the performance of all other firms in relation to this best achieved performance. Thus, we are talking of only relative efficiencies, not absolute efficiencies.

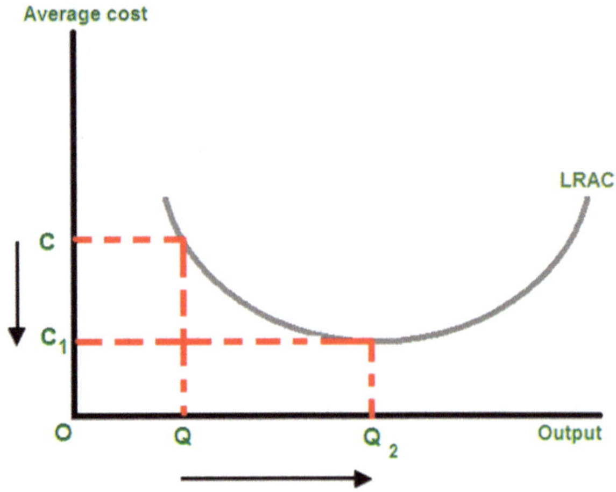

Fig. 9.1 Increase in output from Q to Q_2 causes a decrease in the average cost of each unit from C to C_1

Such an analysis, using efficiency frontier, is often termed as frontier analysis. This efficiency frontier forms the basis of the efficiency analysis. The efficiency frontier envelops the available data, hence the name Data Envelopment Analysis (DEA).

Consider the DMU which does not lie on the frontier. This DMU is inefficient. The following question arises: Can we make a quantitative estimate of its efficiency in relation to the performance of the best firm lying on the frontier?

9.2.1.3 Economies of Scale (ES)

The increase in efficiency of production as the number of goods being produced increases is known as economies of scale. Typically, an agricultural company that achieves economies of scale lowers the average cost per unit through increased production since fixed costs are shared over an increased number of goods.

Economies of scale means that as a company grows and production units increase, the company will have a better chance to decrease its costs. In Fig. 9.1, a generic Long-Run Average Cost (LRAC) curve is represented to illustrate the concept.

Economies of scale are the cost advantages that a firm obtains due to expansion. This should not be confused with increasing returns to scales where simply increasing output within current capacity reduces the short-run cost per unit.

Figure 9.1 shows a simple example, and in real life, there are countering forces of diseconomies of scale. Diseconomies of Scale (DS) are the forces that cause larger firms to produce goods and services at increased per unit costs. As these forces balance, an optimum production volume can be found referred to as constant returns to scales.

Economies of scale refers to the decreased per unit cost as output increases. More clearly, the initial investment of capital is spread over an increasing number of units of output, and therefore, the marginal cost of producing a good or service decreases as production increases (note that this is only in an industry that is experiencing economies of scale).

As we mentioned before, diseconomies may also occur. They could stem from inefficient managerial or labour policies, over-hiring or deteriorating transportation networks (external DS). Furthermore, as a company's scope increases, it may have to distribute its goods and services in progressively more dispersed areas. This can actually increase average costs resulting in diseconomies of scale.

Some efficiencies and inefficiencies are more location specific, while others are not affected by area. If a company has many plants throughout a country, they can all benefit from costly inputs such as advertising. However, efficiencies and inefficiencies can alternatively stem from a particular location, such as a good or bad climate for farming. When ES or DS are location specific, trade is used in order to gain access to the efficiencies.

The key to understanding economies of scale and diseconomies of scale is that the sources vary. A company needs to determine the net effect of its decisions affecting its efficiency and not just focus on one particular source. Thus, while a decision to increase its scale of operations may result in decreasing the average cost of inputs (volume discounts), it could also give rise to diseconomies of scale if its subsequently widened distribution network is inefficient because not enough transport trucks were invested in as well. Thus, when making a strategic decision to expand, companies need to balance the effects of different sources of economies of scale and diseconomies of scale so that the average cost of all decisions made is lower, resulting in greater efficiency all around.

9.2.1.4 Returns to Scales

Refers to a technical property of production that examines changes in output subsequent to a proportional change in all inputs (where all inputs increase by a constant factor). If output increases by that same proportional change, then there are constant returns to scales (CRS). If output increases by less than that proportional change, there are decreasing returns to scales (DRS). If output increases by more than that proportion, there are increasing returns to scales (IRS).

As a short example, where all inputs increase by a factor of 2, new values for output should be:

- Twice the previous output given = a constant returns to scales (CRS)
- Less than twice the previous output given = a decreased returns to scales (DRS)
- More than twice the previous output given = an increased returns to scales (IRS)

9.2.1.5 Allocative Efficiency

Allocative efficiency is a situation in which the limited resources of a firm are allocated in accordance with the wishes of consumers. An allocatively efficient economy produces an "optimal mix" of commodities.

A firm is allocatively efficient when its price is equal to its marginal costs in a perfect market.

Allocative efficiency means efficient distribution of resources: an economic situation where no possible reorganisation of production resources can make some consumers better off without making other consumers worse off.

If price information is available and a behaviour objective is appropriate, then it is possible to measure allocative efficiencies as well as technical efficiencies. Behaviour objectives could be cost minimisation or revenue or profit maximisation. Cost minimisation and revenue maximisation together imply profit maximisation.

9.2.1.6 Factors Which Could Influence the Efficiency of a Farm

These factors are not traditional inputs and are assumed not under the control of the manager. Some examples are:

- Ownership differences (public/private, corporate/noncorporate)
- Coal-fired electric power station influenced by coal quality
- Electric power distribution networks influenced by population density and average customer size
- Schools influenced by socio-economic status of children and city/country location
- Labour union power
- Government regulations

9.2.2 DEA Models

As we mentioned above, the organisational units and farms are more generally called Decision-Making Units (DMUs). DMUs can also be manufacturing units, departments of a big organisation such as universities, schools, bank branches, hospitals, medical practitioners, power plants, police stations, tax offices, prisons, defence bases or a set of firms. In the area of tourism, DMUs can be hotels, motels, destinations, tourism websites and so on.

Efficiency of a decision-making unit is defined as the ratio between a weighted sum of its outputs and a weighted sum of its inputs. We can find the DMU (or the DMUs) having the highest ratio. We call it DMU_o. Then we can compare the performance of all other DMUs relative to the performance of DMU_o. We can calculate the relative efficiency of the DMUs.

Suppose there are n DMUs, $DMU_j, j = 1, 2, \ldots, n$. Suppose m input items and s output items are selected:

- Let the input data for DMUs be $\underset{m\times n}{X} = \left(x_{ij}\right)_{i=1,\ldots,m;j=1,\ldots,n}$.
- Let the output data for DMUs be $\underset{s\times n}{Y} = \left(y_{kj}\right)_{k=1,\ldots,s;j=1,\ldots,n}$.

Given the data, we can measure the efficiency of each DMU$_j$, $j = 1, 2, \ldots, n$. Hence, we need n optimisations (one for each DMU to be evaluated).

Let the DMU we are evaluating be designated as DMU$_o$ ($o = 1, 2, \ldots, n$).

9.2.2.1 Charnes, Cooper and Rhodes (CCR) Model

We will define the CCR-efficiency taking into account all input excesses and output shortfalls. The input-oriented CCR model aims to minimise inputs while satisfying at least the given output levels. The output-oriented CCR model attempts to maximise outputs without requiring more of any of the observed input variables.

Based on the matrix (X, Y), where X is an ($m \times n$) matrix and Y is an ($s \times n$) matrix, the envelopment form of the CCR model is expressed as follows:

$$\min_{\theta,\lambda} \theta \tag{9.1}$$

subject to $\theta x_o - X\lambda \geq 0$, $Y\lambda \geq y_o$ and $\lambda \geq 0$ where, for any DMU$_o$, $\underset{m\times 1}{x_o} = (x_{1o}, x_{2o}, \ldots, x_{mo})^{\mathrm{T}}$, θ is a real variable and $\underset{n\times 1}{\lambda} = (\lambda_1, \ldots, \lambda_n)^{\mathrm{T}}$ is a non-negative vector.

For all DMUs, together we have the following matrix notations:

$$\underset{n\times 1}{\theta}, \underset{n\times n}{\lambda} = \left(\lambda_{jj}\right)_{j=1,\ldots,n;j=1,\ldots,n} \text{and} \min_{\theta,\lambda} \underset{1\times n}{\theta} \tag{9.2}$$

subject to $\underset{m\times 1}{x_o} \underset{1\times n}{\theta} - \underset{m\times n}{X} \underset{n\times n}{\lambda} \geq \underset{m\times n}{0}, \underset{s\times n}{Y} \underset{n\times 1}{\lambda} \geq \underset{s\times 1}{y_o}$ and $\underset{n\times 1}{\lambda} \geq \underset{n\times 1}{0}$

The optimal θ is denoted by θ^*. It is greater than zero and not greater than 1, or $0 < \theta^* \leq 1$.

We define slack vectors by $\underset{m\times 1}{s^-} = \underset{m\times 1}{x_o} \underset{1\times 1}{\theta} - \underset{m\times n}{X} \underset{n\times 1}{\lambda}$ and $\underset{s\times 1}{s^+} = \underset{s\times n}{Y} \underset{n\times 1}{\lambda} - \underset{s\times 1}{y_o}$.

Definition (CCR-efficiency): If an optimal solution (θ^*, λ^*, s^{-*}, s^{+*}) of the CCR model satisfies $\theta* = 1$, $s^{-*} = 0$ and $s^{+*} = 0$, then the DMU$_o$ is called CCR-efficient. Otherwise, the DMU$_o$ is called CCR-inefficient.

The condition $\theta^* = 1$ is referred to as "radial efficiency". The term "weak efficiency" is sometimes used when attention is restricted to the condition $\theta^* = 1$ (also called "Farrell efficiency"). The conditions $\theta^* = 1$, $s^{-*} = 0$ and $s^{+*} = 0$, taken together, describe what is also called "Pareto-Koopmans" or "strong" efficiency.

Definition (Pareto-Koopmans efficiency): A DMU is fully efficient if and only if it is not possible to improve any input or output without worsening some other input or output.

Definition (reference set): For an inefficient DMU_o, we define its reference set E_o by $E_o = \{j \mid \lambda j^* > 0\}, j = 1, \ldots, n$.

An optimal solution can be expressed as

$$\underset{m\times 1}{x_o}\ \underset{1\times 1}{\theta^*} = \underset{m\times n}{X}\ \underset{n\times 1}{\lambda^*} + \underset{m\times 1}{s^{-*}} = \sum_{j\in E_o} x_j\lambda_j^* + s^{-*}$$

$$\underset{s\times 1}{y_o} = \underset{s\times n}{Y}\ \underset{n\times 1}{\lambda}^* - \underset{s\times 1}{s^{+*}} = \sum_{j\in E_o} y_j\lambda_j^* - s^{+*} \tag{9.3}$$

The efficiency of (x_o, y_o) for DMU_o can be improved by the formula

$$\underset{m\times 1}{\hat{x}_o} = \underset{m\times 1}{x_o}\ \underset{1\times 1}{\theta^*} - \underset{m\times 1}{s^{-*}} \le \underset{m\times 1}{x_o}$$

$$\underset{s\times 1}{\hat{y}_o} = \underset{s\times 1}{y_o} + \underset{s\times 1}{s^{+*}} \ge \underset{s\times 1}{y_o} \tag{9.4}$$

This formula for improvement is called the CCR-projection.

Theorem: *The improved activity* $(\hat{x}_o, \hat{y}_o)$ *defined by the CCR-projection is CCR-efficient.*

Corollary to theorem: *The point with coordinates* $(\hat{x}_o, \hat{y}_o)$

$$\underset{m\times 1}{\hat{x}_o} = \underset{m\times 1}{x_o}\ \underset{1\times 1}{\theta^*} - \underset{m\times 1}{s^{-*}} = \sum_{j\in E_o} x_j\lambda_j^*$$

$$\underset{s\times 1}{\hat{y}_o} = \underset{s\times 1}{y_o} + \underset{s\times 1}{s^{+*}} = \sum_{j\in E_o} y_j\lambda_j^* \tag{9.5}$$

is the point on the efficient frontier used to evaluate the performance of DMU_o.

9.2.2.2 The Output-Oriented CCR Model

The output-oriented CCR model attempts to maximise outputs while using no more than the observed amount of any input.

The slack (t^-, t^+) of the output-oriented model is defined by

$$t^- = x_o - X\mu$$

$$t^+ = Y\mu - \eta y_o \tag{9.6}$$

η^* satisfies $\eta^* \geq 1$. The higher the value of η^*, the less efficient the DMU is. η^* expresses the output enlargement rate.

An input-oriented CCR model is efficient for any DMU if and only if it is also efficient when the output-oriented CCR model is used to evaluate its performance. The solution of the output-oriented CCR model may be obtained from that of the input-oriented CCR model.

The improvement using output-oriented CCR model is expressed by

$$\hat{x}_o = x_o - t^{-*} = \sum_{j \in E_o} x_j \mu_j^*$$
$$\hat{y}_o = \eta^* y_o + t^{+*} = \sum_{j \in E_o} y_j \eta_j^* \tag{9.7}$$

9.2.2.3 Banker, Charnes and Cooper (BCC) Model

The BCC problem is solved using a two-phase procedure. In the first phase, we minimise θ_B, and, in the second phase, we maximise the sum of the input excesses and output shortfalls, keeping $\theta_B = \theta_B{}^*$. Here $\theta_B{}^*$ is the optimal value obtained in the first phase. An optimal BCC solution is represented by $(\theta_B{}^*, \lambda^*, s^{-*}, s^{+*})$, where s^{-*} and s^{+*} represent the maximal input excesses and output shortfalls, respectively.

Definition (BCC-efficiency): If an optimal BCC solution $(\theta_B{}^*, \lambda^*, s^{-*}, s^{+*})$ satisfies $\theta_B{}^* = 1$, $s^{-*} = 0$ and $s^{+*} = 0$, then the DMU_o is called BCC-efficient.

We have the following formula for improvement:

$$\hat{x}_o = \theta_B * x_o - s^{-*} - *, \hat{y}_o = y_o + s^{+*} \tag{9.8}$$

Theorem: *The improved activity $(\hat{x}_o, \hat{y}_o)$ is BCC-efficient.*

Theorem: *A DMU that has a minimum input value for any input item, or a maximum output value for any output item, is BCC-efficient.*

9.2.2.4 The Increasing Returns to Scales Model (IRS) and the Decreasing Returns to Scales Model (DRS) or Relaxation of the Convexity Condition

The BCC envelopment model can be extended by relaxing the convexity condition $e\lambda = 1$ to $L \leq e\lambda \leq U$, where L, $(0 \leq L \leq 1)$ and U, $(1 \leq U)$ are lower and upper bounds for the sum of the λ_j. Notice that $L = 0$, $U = \infty$ corresponds to the CCR model and $L = U = 1$ corresponds to the BCC model. Two typical extensions are discussed below.

The case $L=1$, $U=\infty$ is called the *Increasing Returns to Scales* (IRS) or Non-Decreasing Returns to Scales (NDRS) model. The case $L=0$, $U=1$ is called the *Decreasing Returns to Scales* (DRS) or the *Non-Increasing Returns to Scales* (NIRS) model.

9.2.2.5 The Increasing Returns to Scales Model (IRS)

The constraint on λ is $e\lambda \geq 1$. The interpretation of this constraint is that we cannot reduce the scale of DMU but it is possible to expand the scale to infinity. The output/input ratio for any point on the efficient frontier is not decreasing with respect to input. The term NDRS is derived from that fact. That is, a proportional increase in output is always at least as great as the related proportional increase in output and is always at least as great as the related proportional increase in input. In mathematical terms, $\Delta y\,/y \geq \Delta x\,/x$, where Δy, Δx are the increases to be made from a frontier point with coordinate (x, y). *This model focuses on the scale efficiencies of relatively small DMUs.*

9.2.2.6 The Decreasing Returns to Scales (DRS) Model

The constraints on λ are $0 \leq e\lambda \leq 1$. The interpretation of these constraints is that scaling up of DMUs is interdicted and scaling down is permitted. The output/input ratio of efficient frontier points is decreasing with respect to the input scale. That is, $\Delta y\,/y \;=\; \Delta x\,/x$ for the first segment on the frontier and strict inequality $\Delta y\,/y \;<\; \Delta x\,/x$ holding thereafter. This model puts the emphasis on larger DMUs where returns to scales is decreasing.

It is logically true that for every DMU we have the relations $\theta^*_{\mathrm{CCR}} \leq \theta^*_{\mathrm{IRS}}, \theta^*_{\mathrm{DRS}} \leq \theta^*_{\mathrm{BCC}}$.

9.2.2.7 Model Sources of Inefficiency

It is interesting to investigate the sources of inefficiency that a DMU might have. Are they caused by the inefficient operation of the DMU itself or by the disadvantageous conditions under which the DMU is operating?

For this purpose, comparisons of the (input-oriented) CCR and BCC scores deserve consideration. The CCR model assumes the constant returns to scales production possibility set. It is postulated that the radial expansion and reduction of all observed DMUs and their non-negative combinations are possible and hence the CCR score is called global technical efficiency. The BCC model assumes that convex combinations of the observed DMUs form the production possibility set and the BCC score is called local pure technical efficiency. If a DMU is fully efficient in both the CCR and BCC scores, it is operating in the most productive scale size.

If a DMU has full BCC-efficiency but a low CCR score, then it is operating locally efficiently but not globally efficiently due to the scale size of DMU. Thus, it is reasonable to characterise the scale efficiency of a DMU by the ration of CCR and BCC scores. We define scale efficiency as follows:

Definition: Let the CCR and BCC scores of a DMU be θ^*_{CCR} and θ^*_{BCC}, respectively. The scale efficiency (SCAL) is defined by

$$\text{SE} = \frac{\theta^*_{\text{CCR}}}{\theta^*_{\text{BCC}}} \tag{9.9}$$

SCAL is not greater than 1. The BCC score expresses the (local) Pure Technical Efficiency (PTE) under variable returns to scales circumstances. The CCR score is called the (global) Technical Efficiency (TE) since it takes no account of scale effect as distinguished from PTE. For a BCC-efficient DMU with constant returns to scales characteristics (i.e. in the most productive scale size), the scale efficiency (SCAL) is 1.

9.2.2.8 SBM Model

We introduce a new measure ρ called SBM (Slacks-Based Measure). It is invariant to the units of measure used for the different inputs and outputs. This new measure is a scalar that yields the same efficiency value when distances are measured in either kilometres or miles. More generally, this measure is the same when x_{io} and x_{ij} are replaced by $k_i x_{io} = \hat{x}_{io}$ and $k_i x_{ij} = \hat{x}_{ij}$ and y_{ro} and y_{rj} are replaced by $c_r y_{ro} = \hat{y}_{ro}$ and $c_r y_{rj} = \hat{y}_{rj}$, where k_i and c_r are positive constants, $i = 1, \ldots, m$, $r = 1, \ldots, s$. This property is known as "units invariant". The SBM measure is monotone decreasing in each input and output slack. This property is known as "monotone".

Slacks-based measure ρ can be interpreted as the ratio of mean input and output mix inefficiencies.

Theorem: *If DMU **A** dominates DMU **B** so that* $x_A \leq x_B$ *and* $y_A \leq y_B$, *then* $\rho^*_A \geq \rho^*_B$.

Definition (SBM-efficient): A DMU (x_o, y_o) is SBM-efficient if and only if $\rho^* = 1$.

This condition is equivalent to $s^{-*} = 0$ and $s^{+*} = 0$, i.e. no input excess and no output shortfall in an optimal solution.

For an SBM-inefficient DMU (x_o, y_o), we have the expression:

$$\begin{aligned} x_o &= X\lambda^* + s^{-*}, \\ y_o &= Y\lambda^* - s^{+*}. \end{aligned} \tag{9.10}$$

The DMU (x_o, y_o) can be improved and becomes efficient by deleting the input excesses and augmenting the output shortfalls. This is accomplished by SBM-projection expressed by the following formulae, called SBM-projection:

$$\begin{aligned} \hat{x}_o &= x_o - s^{-*}, \\ \hat{y}_o &= y_o + s^{+*}. \end{aligned} \tag{9.11}$$

which are the same as for the additive model.

We will define the reference set for (x_o, y_o) as the following:
Definition (reference set): The set of indices R_o corresponding to positive λ_j^*s is called the reference set for (x_o, y_o).

Using the reference set R_o, we can express $(\hat{x}_o, \hat{y}_o)$ by

$$\hat{x}_o = \sum_{j \in R_o} x_j \lambda_j^*$$

$$\hat{y}_o = \sum_{j \in R_o} y_j \lambda_j^* \tag{9.12}$$

This means that the point on the efficient frontier $(\hat{x}_o, \hat{y}_o)$ is expressed as a positive combination of the members of the reference set R_o. The members of the reference set R_o are also efficient.

Theorem: *The optimal SMB ρ^* is not greater than the optimal CCR θ^*.*

This theorem reflects the fact that SBM accounts for *all inefficiencies* whereas θ^* accounts only for *"purely technical" inefficiencies.*

The relation between CCR-efficiency and SMB-efficiency is given in the following theorem:

Theorem: *A DMU (x_o, y_o) is CCR-efficient if and only if it is SMB-efficient.*

9.2.2.9 Outlier Detection in PAR

The main drawback of deterministic frontier models is that they are very sensitive to outliers and extreme values and that noisy data are not allowed. We perform outlier analysis using the method described in Wilson (1993). This chapter describes an influence-function approach for detecting outliers in the context of frontier models.

The graphic analysis based on outlier statistic developed in Wilson (1993) and implemented in FEAR is used to identify observations in DEA models that are possible outliers. A line in the log-ratio plot connects the second smallest value of the ratios for each observation deleted to illustrate the separation between the smallest ratios for each observation. The plot is approximately linear under the homogeneity model. Under the heterogeneity model, the log-ratio plot shows convexity.

9.2.2.10 Some Notes on CCA and Some Related Methods

Canonical correlation analysis (CCA) is a multidimensional exploratory statistical method.

A canonical correlation is the correlation of two latent (canonical) variables, one representing a set of independent variables, the other a set of dependent variables.

Each set may be considered a latent variable based on measured original variables in its set. The canonical correlation is optimised such that the linear correlation between the two latent variables (called canonical variates) is maximised. There may be more canonical variates relating the two sets of variables. The purpose of canonical correlation is to explain the relation of the two sets of variables, not to model the individual variables. For each canonical variate, we can also assess how strongly it is related to measured variables in its own set or the set for the other canonical variate.

Both methods, principal components analysis (PCA) and CCA, have the same mathematical background. The main purpose of CCA is the exploration of sample correlations between two sets of quantitative variables, whereas PCA deals with one data set in order to reduce dimensionality through linear combination of initial variables.

Another well-known method can deal with the same kind of data: Partial Least Squares (PLS) regression. However, the object of PLS regression is to explain one or several response variables (outputs) in one set by way of variables in the other one (the input). On the other hand, the object of CCA is to explore correlations between two sets of variables whose roles in the analysis are strictly symmetric. As a consequence, mathematical principles of both PLS and CCA methods are fairly different.

9.2.2.11 Variable Aggregation in PAR

The question of obtaining an appropriate aggregate input from appropriate individual inputs is an important one. A natural way to define an aggregate input is to assume a linear structure of aggregation of the input variables. One of the most important issues here is the choice of weights in the aggregation.

A natural extension of the aggregation of input or output techniques is the use of weight restrictions. The use of weight restrictions is a much more subtle technique. For example, instead of eliminating an unimportant input or output, which is the same as assigning a zero weight to it, we may restrict its weight to be low in relation to the more important inputs and outputs. This way the unimportant parameter will still count in the overall model but only up to the specified limit of "importance".

Weight choice may be done by the researcher according to his opinion about the contribution of each variable. In our approach, we use Canonical Correlation Analysis (CCA) to aggregate automatically both input and output data sets.

Obviously the input and output sets of variables in a production process are related. We are concerned with determining a relationship between the two sets of variables. The aim is the linear combinations that maximise the canonical correlation to be found. Such a linear combination is called *canonical variate*.

In this chapter, we propose CCA to aggregate both input and output variables to get final input and output, respectively.

The aggregation in PAR approach is not fixed, and because of it, we are giving the answer of the following two important questions that arise frequently.

9.2.2.12 Variable Selection in PAR

Variable selection in DEA is problematic. The estimated efficiency for any DMU depends on the inputs and outputs included in the model. It also depends on the number of outputs plus inputs. It is clearly important to select parsimonious specifications and to avoid as far as possible models that assign full high efficiency ratings to DMUs that operate in unusual ways.

In practice, when we apply DEA, the number of DMUs should be greater than the total amount of variables in both sets. Usually in real-world applications, the number of DMUs is restricted. Because of it, one of the most important steps in the modelling using DEA is the choice of input and output variables.

Variable selection is crucial to the process as the omission of some of the inputs can have a large effect on the measure of efficiency. It is now recognised that improper variable selection often results in biased DEA evaluation results.

The attention to variable selection is particularly crucial since the greater the number of input and output variable, the less discerning are the DEA results (Jenkins and Anderson 2003). However, there is no consensus on how best to limit the number of variables.

Several methods have been proposed that involve the analysis of correlation among the variables, with the goal of choosing a set of variables that are not highly correlated with one another. Unfortunately, studies have shown that these approaches yield results which are often inconsistent in the sense that removing variables that are highly correlated with others can still have a large effect on the DEA results (see Nunamaker 1985). Other approaches look at the change in the efficiencies themselves as variables are added and removed from the DEA models, often with a focus on determining when the changes in the efficiencies can be considered statistically significant. As part of these approaches, procedures for the selection of variables to be included in the model have been developed by sequentially applying statistical techniques.

Another commonly used approach for reducing the list of variables for inclusion in the DEA model is to apply regression and correlation analysis (Lewin et al. 1982). This approach purports those variables which are highly correlated with existing model variables are merely redundant and should be omitted from further analysis. Therefore, a parsimonious model typically shows generally low correlations among the input and output variables, respectively, Chilingerian (1995) and Salinas-Jimenez and Smith (1996).

The authors Norman and Stoker (1991) noted that the observation of high statistical correlation alone was not sufficient. A logical causal relationship to explain why the variable influenced performance was necessary. Another application of variable selection based on correlating the efficiency scores can be found in Sigala et al. (2004).

In this chapter, we propose CCA to be used in order for the most appropriate variables to be selected. In PAR approach, we apply CCA to select both input and output variables and to get final input and output sets, respectively.

9.3 Azorean Farms' Efficiency Measurement

The Azores islands belong to the Portuguese territory with a population of about 250,000 inhabitants. The main economic activity is dairy and meat farming. Dairy policy depends on Common Agricultural Policy of the European Union and is limited by quotas. In this context, decision makers need knowledge for deciding the best policies in promoting quality and best practices. One of the goals of our work is to provide Azorean government with a reliable tool for measurement of productive efficiency of the farms.

The names of all input variables used in analysis are the following: EquipmentRepair, Oil, Lubricant, EquipmentAmortization, AnimalConcentrate, VeterinaryAndMedicine, OtherAnimalCosts, PlantsSeeds, Fertilizers, Herbicides, LandRent, Insurance, MilkSubsidy, MaizeSubsidy, SubsidyPOSEIMA, AreaDimension and DairyCows. The names of output variables are Milk and Cattle.

We start the data analysis with outlier detection. One outlier obtained in Terceira data was the result of a recording error and was corrected. We used again the statistical methodology presented in Wilson (1993) and implemented it in FEAR package to look for new atypical observations. Using the graphical analysis presented in Fig. 9.2, another observation could also be identified as an outlier. However, data from Terceira Island is viewed as coming from a probability distribution, and it is quite possible to observe one point with low probability. One would not expect to observe many such points, given their low probability. The fact that a particular observation has low probability of occurrence is not sufficient to warrant the conclusion that this observation is an error. More errors in the available data are not identified.

Canonical correlation analysis aims at highlighting correlations between input and output data sets. Two preliminary steps calculate the sample correlation coeffi-

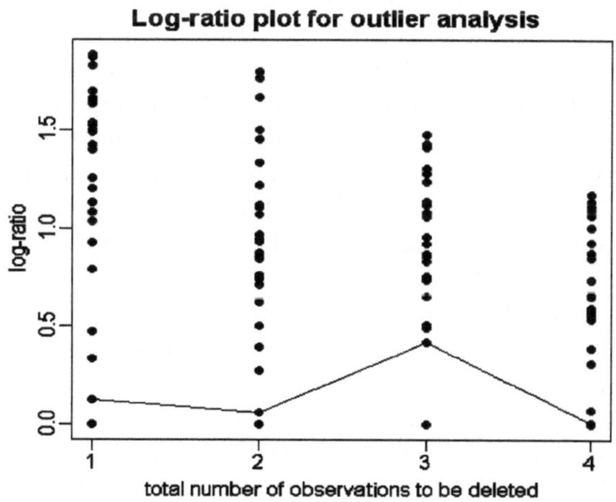

Fig. 9.2 Plot produced by the outlier detection procedure

Table 9.1 Sample correlation coefficients

	Milk	Cattle
EquipmentRepair	0.399089550	0.449336923
Oil	0.349190515	−0.023206764
Lubricant	0.009272362	−0.171455723
EquipmentAmortization	0.051043354	−0.077088336
AnimalConcentrate	0.914685924	0.537983929
VeterinaryAndMedicine	0.707943660	0.370392398
OtherAnimalCosts	0.724266952	0.407358115
PlantsSeeds	0.719946680	0.304399253
Fertilizers	0.781448807	0.452145566
Herbicides	0.497643020	0.347245965
LandRent	0.722516988	0.343699321
Insurance	−0.072519332	0.002379461
MilkSubsidy	0.746508776	0.431464776
MaizeSubsidy	0.751413121	0.526768325
SubsidyPOSEIMA	0.724407535	0.083726114
AreaDimension	0.536678292	0.279164537
DairyCows	0.776032879	0.348513730

cients and visualise the correlation matrixes. All sample correlation coefficients are presented in Table 9.1.

This table highlights a significant correlation between Milk and AnimalConcentrate and nearly null correlation between Milk and Lubricant, Milk and EquipmentAmortization and Milk and Insurance.

In practice, the number of DMUs should be greater than the total amount of variables in both input and output sets. Any resource used by an Azorean dairy farm is treated as an input variable, and because of it, the list of variables that provide an accurate description of the milk and meat production process is large.

This example is focused on measuring efficiency when the number of DMUs is few and the number of explanatory variables needed to compute the measure of efficiency is too large. We approach this problem from a statistical standpoint through both variable selection and variable aggregation approaches.

The results from CCA are printed in the following table.

From Table 9.2, we can conclude that both canonical variates are predominantly associated with the original inputs AnimalConcentrate and Fertilizers and with the original output variable Milk. In this way, we select the two input variables AnimalConcentrate and Fertilizers and one output variable Milk.

On Fig. 9.3, the input and output variables are plotted on the first two canonical variates. Variables with a strong relation are projected in the same direction from the origin. The greater the distance from the origin, the stronger the relation is. The following variables, AnimalConcentrate, VeterinaryAndMedicine, OtherAnimalCosts, MilkSubsidy, MaizeSubsidy, Herbicides, Fertilizers, PlantsSeeds, LandRent, AreaDimension, DairyCows and Milk, are a set of variables with a stronger relation than the rest. In this set, AnimalConcentrate, DairyCows, VeterinaryAndMedicine, OtherAnimalCosts and MilkSubsidy are the variables with the most strong relation. MaizeSubsidy and Herbicides are also variables with a strong relation.

Table 9.2 Correlations of the original outputs with both aggregated input and output

	$scores$corr.*Y*.*x*scores	$scores$corr.*Y*.*y*scores
Milk	−0.9529591	−0.9953781
Cattle	−0.5225409	−0.5458007
	$scores$corr.*X*.*x*scores	$scores$corr.*X*.*y*scores
EquipmentRepair	−0.44487248	−0.42591381
Oil	−0.34213524	−0.32755482
Lubricant	0.01024649	0.00980983
EquipmentAmortization	−0.04167289	−0.03989696
AnimalConcentrate	−0.96395974	−0.92287966
VeterinaryAndMedicine	−0.74087590	−0.70930276
OtherAnimalCosts	−0.76117503	−0.72873682
PlantsSeeds	−0.74525915	−0.71349921
Fertilizers	−0.82269954	−0.78763940
Herbicides	−0.53062365	−0.50801061
LandRent	−0.75224389	−0.72018629
Insurance	0.07133021	0.06829041
MilkSubsidy	−0.78586254	−0.75237225
MaizeSubsidy	−0.80148885	−0.76733263
SubsidyPOSEIMA	−0.72469294	−0.69380945
AreaDimension	−0.56145996	−0.53753280
DairyCows	−0.80562574	−0.77129323

Both the original inputs and outputs are aggregated into overall measures called aggregate input variate and aggregate output variate.

Then we use aggregated input and output in DEA formulation.

We build the DEA analysis on aggregated measures. On Fig. 9.4, all DMUs and the efficient frontier are visualised.

9.4 Conclusions

PAR (Productivity Analysis with R) is implemented in R statistical software version 2.8.1 using the DEA, FEAR and CCA packages and routines developed by us (see R Development Core Team, 2007). PAR is a very flexible, extensible software based on CCA and DEA models, implemented as CCA and FEAR packages in R. The cost of this flexibility is that the user must type commands at a command-line prompt.

In PAR methodology, CCA provides an aggregation of both input and output units and then DEA provides efficient units. The aggregation can cause significant additional bias in a DMU's technical efficiency scores. The effects of the input aggregation on efficiency indicators have been investigated. This study used data from Terceira Island. Azorean government can apply our approach to other islands and to find "the best practice" of Azorean agricultural system.

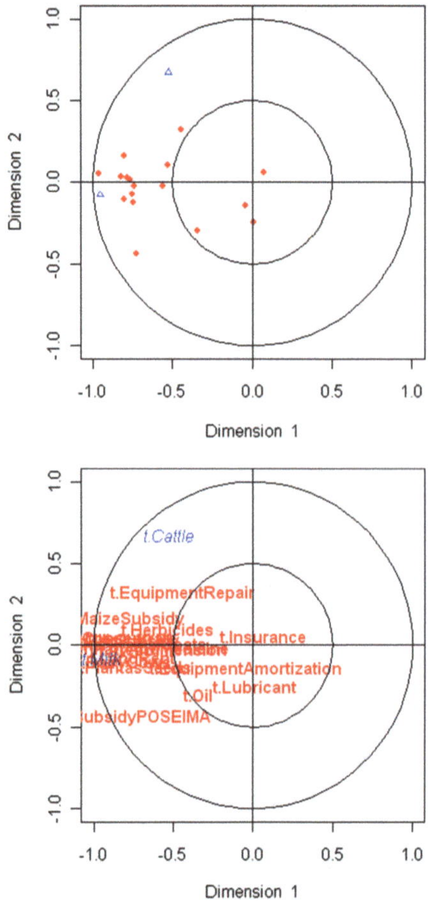

Fig. 9.3 Input and output variables plotted on the first two canonical variates

In spite of the good results achieved, it is important to recognise the major limitations and possible problems in conducting a DEA:

- Measurement error and other noise may influence the shape and position of the frontier.
- Outliers may influence the results. Because of it, we always start with outlier detection.
- The exclusion of an important input or output can result in biased results. Because of it, a variable aggregation method is proposed by PAR.
- The efficiency scores obtained are only relative to the best firms in the sample. The inclusion of extra firms (e.g. from overseas) may reduce efficiency scores.
- Be careful when comparing the mean efficiency scores from two studies. They say nothing about the efficiency of one sample relative to the other.
- The addition of an extra firm in a DEA analysis cannot result in an increase in the technical efficiency scores of the existing firms.

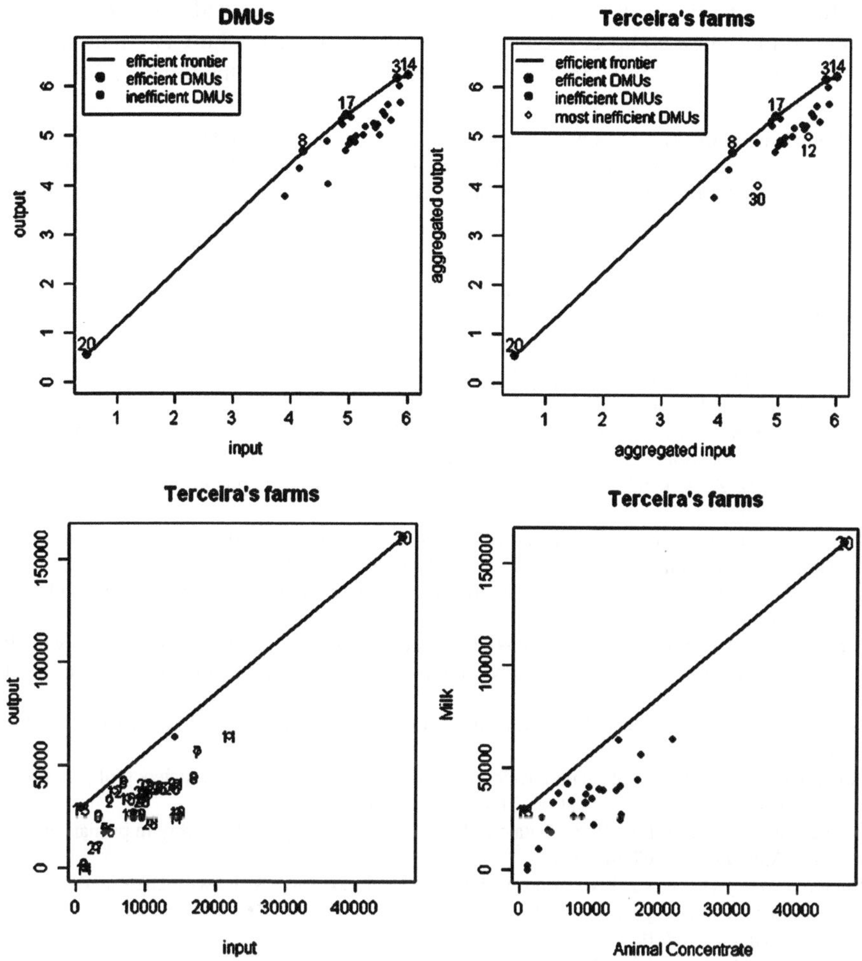

Fig. 9.4 Several examples with and without aggregation using BCC model (first two) and CCR model

- The addition of an extra input or output in a DEA model cannot result in a reduction in the technical efficiency scores.
- With few observations and many inputs and/or outputs, many of the firms will appear on the DEA frontier. If an investigator wishes to make an industry look good, he could reduce the sample size and increase the number of inputs and outputs in order to increase the technical efficiency scores. Because of it, a variable selection method is proposed by PAR.
- Treating inputs and outputs as homogeneous commodities when they are heterogeneous may bias results.

In future work, we are going to use PAR with both real and simulated data in order to find out a compromise between environment, agriculture and tourism and to investigate the potential impact of agricultural tourism on the farms' efficiency.

Acknowledgments This work has been partially supported by Regional Division for Science and Technology of Azores government through the project M.2.1.2/I/009/2008, "Productivity Analysis of Azorean Cattle-Breeding Farms with R Statistical Software".

References

Chilingerian JA (1995) Evaluating physician efficiency in hospitals: a multivariate analysis of best practices. Eur J Oper Res 80:548–574

Cooper WW, Seiford LM, Tone K (2007) Data envelopment analysis: a comprehensive text with models, applications, references and DEA-solver software, 2nd edn. Springer, New York

Diaz-Martinez Z, Fernandez-Menendez J (2008) DEA: data envelopment analysis. R package version 0.1–2

Jenkins L, Anderson M (2003) A multivariate statistical approach to reducing the number of variables in data envelopment analysis. Eur J Oper Res 147:51–61

Lewin AY, Morey RC, Cook TJ (1982) Evaluating the administrative efficiency of courts. Omega 10(4):401–411

Mendes A, Noncheva V, Silva E (2009) Decision support for enhanced productivity with R software: an Azorean farms case study. Thirty eighth annual meeting of WDSI, Hawaii, 7–11 Apr 2009

Norman M, Stoker B (1991) Data envelopment analysis: the assessment of performance. Wiley, Chichester

Nunamaker TR (1985) Using data envelopment analysis to measure the efficiency of non-profit organizations: a critical evaluation. Manag Decis Econ 6(1):50–58

Salinas-Jimenez J, Smith P (1996) Data envelopment analysis applied to quality in primary health care. Ann Oper Res 67:141–161

Sigala M, Airey D, Jones P, Lockwood A (2004) ICT paradox lost? A stepwise DEA methodology to evaluate technology investments in tourism settings. J Travel Res 43:180–192

Simar L, Wilson PW (2000) A general methodology for bootstrapping in non-parametric frontier models. J Appl Stat 27:779–802

Wilson PW (1993) Detecting outliers in deterministic nonparametric frontier models with multiple outputs. J Bus Econ Stat 11:319–323

Wilson PW (2008) FEAR: a software package for frontier efficiency analysis with R. Socio-Econ Plan Sci 42(4):247–254

Chapter 10
Sustainable Tourism and Agriculture Multifunctionality by PAR: A Variable Selection Approach

Armando B. Mendes, Veska Noncheva, and Emiliana Silva

Abstract Data Envelopment Analysis (DEA) is a popular non-parametric method used to measure efficiency. It uses linear programming to identify points on a convex hull defined by the inputs and outputs of the most efficient Decision Making Units (DMUs). Two critical elements account for the strength of the DEA approach: (1) no a priori structure is placed on the production process of the firm, and (2) the models can yield a measure of efficiency even with a very small number of data points. The first point is particularly important because the measure of efficiency is based upon the best practice of the DMUs at any of the levels of output observed.

Data envelopment analysis measures efficiency and is very sensitive to the choice of variables for two reasons: the number of efficient DMUs is directly related to the number of variables, and the selection of the variables greatly affects the measure of efficiency when the number of DMUs is few and/or when the number of explanatory variables needed to compute the measure of efficiency is too large. Our approach advises which variables should be included in a DEA model. Hence, a variable selection method is presented for the deterministic DEA approach. First, a definition of different measures of efficiency and the various DEA models used to measure efficiency is provided, and then a variable selection method is proposed. The Azorean agricultural system is used as an example to illustrate the method.

A.B. Mendes (✉)
CEEAplA, Departamento de Matemática, Universidade dos Açores, Rua da Mãe de Deus, 9500-801 Ponta Delgada, Açores, Portugal
e-mail: amendes@uac.pt

V. Noncheva
CEEAplA, Faculty of Mathematics and Informatics, University of Plovdiv, "PaisiiHilendarski" Plovdiv, Bulgaria

E. Silva
CEEAplA, Departamento de Ciências Agrárias, Universidade dos Açores, Rua Capitão João de Ávila, 9700-042 Angra do Heroísmo, Açores, Portugal
e-mail: emiliana@uac.pt

A.B. Mendes et al. (eds.), *Efficiency Measures in the Agricultural Sector: With Applications*, DOI 10.1007/978-94-007-5739-4_10,

Keywords Data envelopment analysis • Productivity analysis with R • Canonical correlation analysis • Variable selection

10.1 Introduction

Tourism is increasing in Azores islands, although in different proportion per island (S. Miguel 47%, Terceira 24.4% and Faial 12.4%). The main argument for marketing has been the nature and its conservation. The green islands can only stay green if it is possible to have a sustainable compromise between environment, agriculture and tourism.

From the SREA (2007a) characterisation of Azorean tourism, tourists were mainly elder, settled and experienced (around 45–54 years old); they come mainly from Portugal mainland, Nordic countries (Denmark, Norwegian, Sweden, Iceland) and Diaspora countries (United States of America and Canada). The most part of the tourists have higher education and a professional activity. They choose the "Azores destiny" mainly to "relax", "business" or "to visit family and friends", and they are attracted by "landscape", "nature" and "exotism" of the islands. The tourists pay, in average, per trip about 1,193€ and they stay in Azores about 9 days. The establishments preferred are "hotel" and "family and friends houses".

In 2007, the Azores had about 82 establishments (distributed for the nine islands) for agricultural tourism: 54.9% were country houses, 23.2% rural tourism, 17% lodging tourism, 3.7% agrotourism and 1.2% village tourism (SREA 2007b).

For increasing the income of agricultural enterprises, European Union had developed the concept of multifunctionality understood as "a characteristics of an activity which produces multiple and interconnects results and effects" (OECD 2001). The functions of agricultural multifunctionality are various such as agricultural, ecological, cohesion, recreational, educational, cultural and residential. In this case, the rural tourism presents as an alternative of farm income (Rodriguez et al. 2004).

The data available shows that Azores have potentiality to this kind of rural establishments. How can it affect the efficiency of dairy farms, the most representative type in Azorean agriculture? If less extensive grazing system were compensated by the increase of country houses, could this services maintain the same income and raise the efficiency? How much must an agricultural unit receive, from tourism income, to compensate the loss of income by extensive grazing system, without becoming inefficient?

Efficiency was initially measured in Azores farms by Silva et al. (2004). They had measured the Azores dairy farms' technical efficiency by applying a non-parametric efficiency analysis to a panel data of 122 dairy farms from the Azores, Portugal, for 1996. The analysis used DEA with constant and variable returns to scales models, with an input-oriented model approach. Two outputs (milk production and subsidies) and three inputs (agricultural area, number of dairy cows and variable and fixed cost) were considered relevant. The results suggest that the average technical efficiency is very low (66.4%) compared with published research data, and only a few (7%) dairy farms were found to be efficient. In fact, the Azores dairy farms must increase their technical efficiency, given that they operate above their resource capacity. The lower efficiency showed that it is possible to produce the same amount of milk while saving approximately 33.6% of resources (or inputs).

The small dimensions (less than 25 ha per farm) may explain this low efficiency in the Azores. The Azorean farms are smaller than farms in New Zealand, Canada or Australia (Jaforullah and Whiteman 1999; Fraser and Cordina 1999; and Cloutier and Rowley 1993). The last researches suggest bigger farms are more efficient. The inefficiency in the Azorean dairy farms seems to be influenced by the great amount of fixed costs spent on agricultural equipment and animal feeding with concentrates.

In 82 milk dairy farms, Marote and Silva (2002) measured the efficiency in 3 years – 1997, 1998 and 1999 – using DEA. About 63.4, 62.2 and 70.7% of farms were efficient in the 1997, 1998 and 1999, respectively. The technical efficiency of Variable Returns to Scales (VRS) was 0.957, 0.951 and 0.960 in the 3 years mentioned. In this period, the efficiency was similar, but there was an increase of the farms' efficiency.

Later, Marote and Silva (2011) analysed the efficiency of 82 farms from 1997 to 1999 in Terceira Island (Azores archipelago) farms, using DEA. They used two models: model I considered two outputs – milk production and subsidies – and nine inputs, including dimension, animals and other variables and fixed costs. Model II considered one output – milk production – and the same nine inputs of model I.

The efficiency of farms does not improve with subsidies. This conclusion was observed by comparing efficiency using or not using subsidies, probably because the farms balanced the lowest subsidy amount with bigger milk production. This work showed that although the subsidies were very important contributors to the farms' income, their influence in efficiency was very small, which means that the efficiency and the number of farms efficient did not increase very much. Probably, the farms will have a greater efficiency if they rationalise the use of feeding and equipment costs.

Comparing this study with others in same conditions, the efficiency measured in Azores was bigger than in other regions, in part caused by the greater number of inputs. As has been shown by Suhariyanto (1999), more use of variables increases the efficiency value.

Silva and Santos (2007) measured the efficiency of 184 farms of Azores in 2002 using a different system production (milk, meat and mix: milk and meat). They used DEA and the results showed that the technical efficiency at constant returns to scales (CRS) was 63.2%, a variable returns to scales (VRS) about 71.4% and scale (SCA) 89.2% in milk production system. In the meat production system, the efficiency was greater than the milk system, a constant returns to scales (CRS – 69.4%; VRS – 82.9%), and smaller in scale efficiency (SCA – 84.2%). In the mix system production, apparently the most efficient system, the values were the biggest (CRS, 89%; VRS, 99.2%; and SCA, 89.8%) of the three systems. The number of the efficient farms was 9.8% in milk system, 11.1% in the meat system and 46.7% in the mix system.

Using a parametric approach, Venâncio and Silva (2004) measured the efficiency by a Stochastic Frontier Production (SFP) for three groups of farms, using Frontier software. The efficiency in the Faial Island (Azores archipelago) farms was higher than 80%, in the three clusters of farms (82, 93.2 and 85.1% for clusters A, B and C). The variables which contributed for inefficiency were subsidies and equipment costs. The most efficient farms were those with land rent, animal sales and bigger farms, such as those Hallam and Machado (1996) observed for the Portuguese case.

One of the most important steps in the modelling using DEA is the choice of input and output variables. Variable selection is crucial to the process as the omission of some of the inputs can have a large effect on the measure of technical efficiency. The practice has been to select the variables by simply choosing the ones that make economic sense. The criteria for the choice of which explanatory variables (inputs or outputs) to include in a DEA model are rarely made explicit.

In this text the selection of the variables that capture most of the relationship between the inputs and outputs is explored for sustainable tourism and agriculture multifunctionality efficiency measures. Because of it, we are interested in the relationship between both the input and output sets of variables, and Canonical Correlation Analysis (CCA) would be the appropriate method of analysis. CCA is a multidimensional exploratory statistical method. More precisely, at first we would like to investigate the following questions:

1. To what extent can the set of two or more output variables be "explained" by the set of two or more input variables?
2. What contribution does a single input or output variable make to the explanatory power of the set of variables to which the variable belongs?

This chapter is focused on measuring efficiency when the number of DMUs is few and when the number of explanatory variables needed to compute the measure of efficiency is too large. Hence, a statistical approach to variable selection for the deterministic DEA models is presented.

10.2 The Efficiency Approach: Data Envelopment Analysis (DEA)

Two approaches are commonly used to measure efficiency: the parametric approach, which relies on statistical techniques to estimate the parameters of a production function, and the non-parametric approach, which compares the observed inputs and outputs of each firm with that of the most performing firms in the information set. The parametric approach has been subject to persistent criticism, centred on two points: the assumption that the production function has the same functional form for all the firms and the fact that econometric estimation of efficiency can produce biased and inconsistent parameter estimates (since an econometric measure of efficiency reflects the average performance and not the best performance).

Data Envelopment Analysis (DEA) is now the most popular method used to measure efficiency. DEA is a non-parametric method, which does not assume any specific production function. Instead, it uses linear programming to identify points on a convex hull defined by the inputs and outputs of the most efficient firms (DMU). Two critical elements account for the strength of the DEA approach: (1) no a priori structure is placed on the production process of the firm, and (2) the models can yield a measure of efficiency even with a very small number of data points. The first point is particularly important because the measure of efficiency is based upon the best practice of the DMUs at any of the levels of output observed.

For a given set of input and output variables, DEA produces a single comprehensive measure of performance called efficiency score. The CCR model (Charnes et al. 1978) formally introduced the linear programming to measure technical efficiency with the assumption of constant returns to scales. In the CCR model, DMUs adjust either their use of inputs or their outputs to reach the production frontier. The BCC model (Banker et al. 1984a, b) removed the assumption of constant returns to scales, and Charnes and Cooper (1985) proposed the additive DEA model, where both inputs and outputs can be adjusted simultaneously. All models use the distance to one of the facets of the production or cost frontier to generate an efficiency index.

This technique has useful applications in many evaluation contexts. The research presented in Chiang et al. (2004) is aimed at measuring hotel performance of International Tourist Hotels (ITHs) in Taiwan by DEA.

DEA can also be used for destination satisfaction management. The study by Sungsoo (2007), DEA Application for the Tourist Satisfaction Management, showed an application of DEA to a tourist destination, Jeju Island, suggesting that DEA was a useful tool to produce important information in managing destination for tourist satisfaction.

The aim for tourism organisations and businesses was to provide more efficient websites in order to gain competitive advantage. The study by Bauernfeind and Mitsche (2008) provides an example of how DEA can be used to assess the website's efficiency of tourism organisations.

The study by Marianna et al. (2004) proposed a way of assessing ICT (the information and communication technologies) productivity in the tourism industry using DEA. The methodology was applied in a data set from the three-star hotel sector in the United Kingdom.

In Gimenez-Garcia et al.'s (2007) study, a three-step data envelopment analysis model was used to reallocate resources in an organisational network. First, the model identified the excess resources of inefficient units and then reallocated these resources and set the output-oriented production goals for efficient units. Finally, the model recalculates improvement targets for the inefficient units based on the revised remaining resources. The procedure was applied to the analysis of 54 restaurant locations belonging to a Spanish fast-food chain. The results showed that originally efficient restaurants can improve their output by an average of 4.20% after a reallocation of inputs and that this reallocation is beneficial for the entire restaurant chain.

DEA makes it possible to identify efficient and inefficient units in a framework where results are considered in their particular context. The units to be assessed should be relatively homogeneous and were originally called Decision-Making Units (DMUs). DEA is an extreme point method and compares each DMU with only the "best" DMUs.

DEA can be a powerful tool when used wisely. A few of the characteristics that make it powerful are:

1. DEA can handle multiple input and multiple output models.
2. DMUs are directly compared against a peer or combination of peers.
3. Inputs and outputs can have very different units. For example, one variable could be in units of lives saved and another could be in units of dollars without requiring an a priori trade-off between the two.
4. Do not need a functional form.

The same characteristics that make DEA a powerful tool can also create problems. An analyst should keep these limitations in mind when choosing whether or not to use DEA:

1. Since DEA is an extreme point technique, noise such as measurement error can cause significant problems.
2. DEA is good at estimating "relative" efficiency of a DMU, but it converges very slowly to "absolute" efficiency. In other words, it can tell how well peers are doing compared to others peers but not compared to a "theoretical maximum".

Variable selection in DEA is problematic. The estimated efficiency for any DMU depends on the number of inputs and outputs included in the model. It also depends on the number of outputs plus inputs. It is clearly important to select parsimonious specifications and to avoid as far as possible models that assign full high efficiency ratings to DMUs that operate in unusual ways.

In practice, when DEA is applied, the number of DMUs should be greater than the total amount of variables in both sets. Usually in real-world applications, the number of DMUs is restricted. Because of it, one of the most important steps in the modelling using DEA is the choice of input and output variables.

The attention to variable selection is particularly crucial since the greater the number of input and output variable, the less discerning are the DEA results (Jenkins and Anderson 2003). However, there is no consensus on how best to limit the number of variables.

In particular, a few researchers, such as Valdmanis (1992) and Hughes and Yaisawarng (2004), discussed the influence of variable selection on DEA results. They calculated efficiency scores by using alternative sets of variables and analysed the sensitivity of DEA efficiency scores. It is now recognised that improper variable selection often results in biased DEA evaluation results. Therefore, the appropriate variable selection is crucial for the successful application of the DEA technique.

The choice of the variable set in DEA is an empirical issue. Inclusion of many variables is not a viable option in DEA. As the number of variables in the DEA model increases, more and more production units become efficient. On the other hand, when relevant variables are omitted, DEA underestimates efficiency, and the effect of this is more severe than when irrelevant variables are included in the DEA model. Lack of a standard-structured approach to variable selection in DEA makes the task of variable selection even more difficult.

Berger and Humphrey (1997) highlighted the difficulty of variable selection when appraising bank performance using DEA. There was no "perfect approach" on the explicit definition and measurement of the banks' input and output. Further, in choosing the variables, there were some restrictions on the type of variables since there is a need for comparable data to minimise possible bias arising from different accounting practices even among the banks that are bounded by federal bank guidelines. Indian banks were no exception.

In their paper "A Statistical Test for Nested Radial DEA Models", Pastor and Ruiz (2002) focused on analysing the marginal role of a given variable, called candidate, with respect to the efficiency measured by means of a DEA model. First, they have defined a new efficiency contribution measure (ECM), which finally compares the efficiency scores of the two radial DEA models differing in the candidate. This can be either one input or one output. Then, based on ECM, they have also approached the problem from a statistical point of view. They have developed a statistical test that allows us to evaluate the significance of the observed efficiency contribution of the candidate. Eventually, solving this test may provide some useful insights in order to decide the incorporation or the deletion of a variable into/from a given DEA model, on the basis of the information supplied by the data. Two procedures for progressive selection of variables were designed by sequentially applying the test: a forward selection and a backward elimination. These can be very helpful in the initial selection of variables when building a radial DEA model.

Several methods have been proposed that involve the analysis of correlation among the variables, with the goal of choosing a set of variables that are not highly correlated with one another. Unfortunately, studies had shown that these approaches yield results which are often inconsistent in the sense that removing variables that are highly correlated with others can still have a large effect on the DEA results (Nunamaker 1985). In his analysis of DEA modelling, Nunamaker found that for selected DMUs, the addition of a highly correlated variable may substantially alter the DEA efficiency scores. He concluded that because a variable was redundant within a regression model did not mean that it was redundant within a DEA model. The existence of high correlation among variables did not necessarily mean that one of the variables could be excluded without changing the subsequent DEA results. Therefore, it would be unwise to rely strictly on regression and correlation analysis as a means of reducing the number of variables. At best, these quantitative techniques could assist in variable reduction. In a similar vein, Golany and Roll (1989) claim that one-at-a-time regression tests on the inputs and outputs should not be regarded as reliable rules for eliminating variables but rather as indicators for a need to examine some of the variables more closely.

Other approaches look at the change in the efficiencies themselves as variables are added and removed from the DEA models, often with a focus on determining when the changes in the efficiencies can be considered statistically significant. As part of these approaches, procedures for the selection of variables to be included in the model have been developed by sequentially applying statistical techniques.

Another commonly used approach for reducing the list of variables for inclusion in the DEA model was to apply regression and correlation analysis (Lewin et al. 1982). This approach purports those variables which were highly correlated with existing model variables. They are merely redundant and should be omitted from further analysis. Therefore, a parsimonious model typically showed generally low correlations among the input and output variables, respectively, Chilingerian (1995) and Salinas-Jimenez and Smith (1996).

One formal procedure is using a "stepwise" approach to variable selection that estimates the change in the efficiencies as variables are added or dropped from the analysis. This method is intended to produce DEA models that include only those variables with the largest impact on the DEA results. Examples showed that stepwise DEA modelling could be used on larger, realistic problems. While a stepwise procedure can inform for the effect of adding and removing variables in a DEA study, the determination of the "best" model to represent any given situation must rely on managerial judgement and knowledge of the operations of the actual situation being represented.

The authors Norman and Stoker (1991) proposed a method of adding variables to the DEA model one at a time. They started with a simple model involving one single output and one single input. Efficiencies for all the DMUs were then calculated. They claimed that high statistical correlation was an indicator that a particular variable influenced performance. A new variable was then added to the DEA model based on the correlation values and incorporated into the measure of efficiency. The process was repeated until no further influential variables remained. They did note that the observation of high statistical correlation alone was not sufficient. A logical causal relationship to explain why the variable influenced performance was necessary. Another application of variable selection based on correlating the efficiency scores can be found in Sigala et al. (2004).

Färe et al. (1988) consider that the basic information provided by the estimated frontier must remain unaffected by a forward selection or a backward elimination of inputs and outputs. They may be helpful when building a radial DEA model to assess efficiency.

Forward procedure is to be used when the analyst starts with a basic model consisting of the set of available variables he/she considers as essential to evaluate efficiency, and there also exists another set of variables that are thought of as possibly relevant to that end. The variable with the largest value of T statistics, if it is statistically significant, enters the model. The algorithm continues until either all variables are in the model or when at a given step the variable with the largest value of T is not statistically significant. The forward algorithm described above implicitly embodies the prior knowledge and experience of the analyst, as it requires an initial selection of the most relevant variables to define the model to start with.

Backward procedure is to be used when the analyst wonders if the specification set of a given DEA model used to evaluate the efficiency can be simplified by eliminating some of the existing variables without significantly affecting the efficiency scores.

In general, it is not recommended that these kinds of automated procedures be used blindly to identify a "best" model because they can never replace professional judgement in the matter field. Nevertheless, they may complement this judgement with information provided by observed data.

In the backward approach, the goal of the method is to remove those variables that do not have significant influence on the efficiency. CCA advises which variable could be removed. The statistical test supports the decision maker to remove the variables. Several statistical tests which can be used to decide the incorporation of a variable into a DEA model have been proposed (Banker 1996). For instance, Brockett and Golany (1996) asserted that the distribution of efficiency scores is generally unknown and is difficult to describe in a low-dimensional parametric model, and they suggest the application of non-parametric statistical techniques based on rank statistics instead of the efficiency ratings themselves. They propose the use of the Mann–Whitney rank test to evaluate the statistical significance of the differences observed in efficiency within a DEA efficiency evaluation framework. See, for example, Simar (1996) for a discussion on some general aspects of the statistical analysis in DEA-type frontier models.

In general, it is possible to conclude that the process starts by selecting a small set of input and output items at the beginning and gradually enlarge the set to observe the effects of the added items. It is desirable that the number of DMUs (n) exceeds the sum of inputs (m) and outputs (s).

A heuristic formulae is

$$n \geq \max\left(ms, 3\left(m + s\right)\right) \tag{10.1}$$

10.3 Productivity Analysis with R (PAR): A Tool for Measuring Efficiency in Azores

In the PAR project, DEA is applied to distinguish between efficient and inefficient observations of performances. Different statistical methods are applied to assist DEA. For example, canonical correlation analysis assists DEA with both variable aggregation and variable selection. PAR methodology is implemented in R. The output of the PAR computer program intends to be self-explanatory. This makes the system appropriate to support public policies. PAR project is designed to provide a bridge from mathematical models to productivity study using R statistical software.

PAR methodology is aimed at:

- Designing a new "data-oriented" methodology for evaluating the performance of Azorean cattle-breeding farm system
- Offering a computer implementation of PAR methodology
- Locating efficiencies and inefficiencies and supporting public policy decisions

A natural measure of performance is a productivity ratio: the ratio of outputs over inputs, where larger values of this ratio are associated with better performance. Performance is a relative concept. For example, the performance of the meat farm in 2008 could be measured relative to its 2007 performance or it could be measured relative to the performance of another farm in 2008. This farm can also analyse the relative performance of units within the farm.

DMUs can also be manufacturing units, departments of a big organisation such as universities, schools, bank branches, hospitals, medical practitioners, power plants, police stations, tax offices, prisons, defence bases or a set of firms. In the area of tourism, DMUs can be hotels, motels, destinations, tourism websites and so on.

Efficiency of a decision-making unit is defined as the ratio between a weighted sum of its outputs and a weighted sum of its inputs. We can find the DMU (or the DMUs) having the highest ratio. We call it DMU_0. Then we can compare the performance of all other DMUs relative to the performance of DMU_0. We can calculate the relative efficiency of the DMUs.

The input-oriented DEA model aims to minimise inputs while satisfying at least the given output levels. The output-oriented DEA model attempts to maximise outputs without requiring more of any of the observed input variables. Dairy policy in Azorean islands depends on Common Agricultural Policy of the European Union and is limited by quotas, in the moment. This is the reason why output-oriented models are not used in this context.

10.4 Canonical Correlation Analysis in Variable Selection

The PAR approach applies CCA to select both input and output variables and to get final input and output sets, respectively. Canonical Correlation Analysis (CCA) is a multidimensional exploratory statistical method. A canonical correlation is the correlation of two latent (canonical) variables, one representing a set of independent variables and the other a set of dependent variables. Each set may be considered a latent variable based on measured original variables in its set. The canonical correlation is optimised such that the linear correlation between the two latent variables (called canonical variates) is maximised.

CCA finds two vectors that maximise the correlation between the linear combinations assuming that vectors a_1 and b_1 are normalised. The resulting variables U_1 and V_1 are called the first canonical variates and ρ_1 is referred as the first canonical correlation.

The canonical correlation is optimised such that the linear correlation between the two latent variables is maximised. Canonical correlation is used for many-to-many relationships. There may be more than one such linear correlation relating the two sets of variables, with each such correlation representing a different dimension by which the input set of variables is related to the output set. We use canonical correlation to explain the relation of the input and output sets of variables. For both input and output canonical variates, we assess how strongly it is related to measured variables in its own set and the set for the other canonical variates.

Wilks' lambda test is used to test the significance of the first canonical correlation. If $p < 0.05$, the two sets of variables are significantly associated by canonical correlation. Likelihood ratio test is a significance test of all sources (not just the first canonical correlation) of linear relationship between the two canonical variables. It is sometimes wrongly used as a test of the significance of the first or another single canonical correlation in a set of such functions.

Canonical correlation squared is the percent of variance in output set explained by input set of variables. In addition to asking how strong the relationship is between two latent variables, canonical correlation is useful in determining how many dimensions are needed to account for that relationship. Canonical correlation finds the linear combination of variables that produces the largest correlation with the second set of variables. This linear combination, or "root", is extracted and the process is repeated for the residual data, with the constraint that the second linear combination of variables must not correlate with the first one. The process is repeated until a successive linear combination is no longer significant.

Canonical correlation is a member of the Multiple General Linear Hypothesis (MLGH) family and shares many of the assumptions of multiple regression such as linearity of relationships, homoscedasticity (same level of relationship for the full range of the data), interval or near-interval data, untruncated variables, proper specification of the model, lack of high multicollinearity and multivariate normality for purposes of hypothesis testing. It also shares with factor analysis the need to impute labels for the canonical variables based on structure correlations, which function as a form of canonical factor loading; researchers may well impute different labels based on the same data.

As with factor analysis, there may be more than one canonical correlation, each representing an orthogonally separate pattern of relationships between the input and output variables. The maximum number of canonical correlations between two sets of variables is the number of variables in the smaller set.

The first canonical correlation is always the one which explains most of the relationship. The canonical correlations are interpreted in the following way: the square of the canonical correlation is the percent of variance in the canonical variate of the output set of variables explained by the canonical variate for input set. Another way to put it is to say that Rc squared is the percent of variance shared by the canonical variates along this dimension. Pooled Rc2 (pooled canonical correlation) is the sum of the squares of all the canonical correlation coefficients, representing all the orthogonal dimensions in the solution by which input and output sets of variables are related. Pooled Rc2 is used to assess the extent to which one set of variables can be predicted or explained by the other set.

The standardised canonical weights are used to assess the relative importance of an individual variable's contributions to a given canonical correlation. The canonical coefficients are the standardised weights in the linear equation of variables which creates the canonical variables. If an independent variable is totally redundant with another independent variable, its partial coefficient (canonical weight) will be zero. Nonetheless, such a variable might have a high correlation with the canonical variable (i.e. a high structure coefficient).

However, Levine (1977) argues against the procedure above on the ground that the canonical coefficients may be subject to multicollinearity, leading to incorrect judgements. Also, because of suppression, a canonical coefficient may even have a different sign compared to the correlation of the original variable with the canonical variable. Therefore, instead, Levine (1977) recommends interpreting the relations of the original variables to a canonical variable in terms of the correlations, which are called structure correlation coefficients, also known as canonical factor loadings, that is, the correlation of canonical variable scores for a given canonical variable with the standardised scores of an original input variable. The table of structure correlations is sometimes called the factor structure. In summary, the canonical weights have to do with the unique contributions of an original variable to the canonical variable, whereas the structure correlations have to do with the simple, overall correlation of the original variable with the canonical variable.

Alpert and Peterson (1972) noted that canonical weights appear more suitable for prediction, while structure coefficients may better explain underlying (although interrelated) constructs. Variables with correlations of 0.3 or above are interpreted as being part of the canonical variable, and those below are not considered part of the canonical variable.

It is well known that because the weights are partial coefficients whereas the canonical factor loadings are not, if a given variable shares variance with other independent variables entered in the linear combination of variables used to create a canonical variable, its weight is computed based on the residual variance it can explain after controlling for these variables. If an independent variable is totally redundant with another independent variable, its canonical weight will be zero. Nonetheless, such a variable might have a high correlation with the canonical variable (i.e. a high structure coefficient). In summary, the canonical weights have to do with the unique contributions of an original variable to the canonical variable, whereas the structure correlations have to do with the simple, overall correlation of the original variable with the canonical variable.

The canonical coefficients are standardised coefficients, and their magnitudes can be compared. However, Levine (1977) argues against this procedure on the ground that the canonical coefficients may be subject to multicollinearity, leading to incorrect judgements. Also, because of suppression, a canonical coefficient may even have a different sign compared to the correlation of the original variable with the canonical variable. Therefore, instead, Levine recommends interpreting the relations of the original variables to a canonical variable in terms of the correlations of the original variables with the canonical variables – that is, by structure coefficients. This is the standard approach.

The CCA assumptions are:

1. Interval level data are assumed.
2. Linearity of relationships is assumed, though there are nonlinear canonical correlation procedures like OVERALS algorithm (Gifi 1990). In the usual form of canonical correlation, however, analysis is performed on the correlation or variance-covariance matrices, which reflect linear relationships. Of course, one

can insert exponentiated or otherwise nonlinearly transformed variables into either measured variable set in canonical correlation.

3. Low multicollinearity: To the extent that the variables within the independent sets of variables are highly correlated, the canonical coefficients will be unstable. The coefficients for some variables may be misleadingly low or even negative because variance has already been explained by other variables.
4. Homoscedasticity and other assumptions of correlation are assumed. The covariates are created based on the correlation matrix, with regression-like assumptions that the degree of correlation is constant along the full range of the variables being correlated.
5. Minimal measurement error is assumed since low reliability attenuates the correlation coefficient. Canonical correlation can also be quite sensitive to missing data.
6. Unrestricted variance: If variance is truncated or restricted due, for instance, to poor sampling, this can also lead to attenuation of the correlation coefficient.
7. Similar underlying distributions are assumed: If two variables come from unlike distributions, their correlation may be well below +1 even when data pairs are matched as perfectly as they can be, while still conforming to the underlying distributions. That is, the larger the difference in the shape of the distribution of the two variables, the more the attenuation of the correlation coefficient. This assumption may well be violated when correlating an interval variable with a dichotomy or even an ordinal variable.
8. Multivariate normality is required for significance testing in canonical correlation. This assumption is violated when dichotomous, dummy and other discrete variables are used. In such situations, where significance testing is not appropriate, researchers may use a resampling method. The central limit theorem demonstrates, however, that for large samples, indices used in significance testing will be normally distributed even when the variables themselves are not normally distributed, and therefore, significance testing may be employed.
9. Non-singularity in the correlation matrix of original variables. This is the problem of perfect multicollinearity: a unique solution cannot be computed if some variables are redundant, thereby approaching perfect correlation with others in the model. A correlation matrix with redundancy is said to be singular or ill conditioned. Data sets based on survey data, in which there are a large number of questions, are more likely to have redundant items.
10. Adequate sample size must exist to reduce the chances of type II error (thinking you don't have something when you do). Stevens (1986) recommends at least 20 times as many cases as variables in the analysis in order to interpret the first canonical correlation only. For two canonical correlations, Barcikowski and Stevens (1975) recommend 40–60 times as many cases as variables.
11. No or few outliers. Outliers can substantially affect canonical correlation coefficients, particularly if sample size is not very large.

We will investigate the relationship between input and output sets of variables. More precisely, we would like to investigate the following questions:

1. To what extent can the set of two or more output variables be "explained" by the set of two or more input variables?
2. What contribution does a single input or output variable make to the explanatory power of the set of variables to which the variable belongs?

10.5 The Example of Azorean Farms' Efficiency

The Azores islands belong to the Portuguese territory with a population of about 250,000 inhabitants; most part (about 75%) of this population is in S. Miguel and Terceira islands.

The main economic activity is dairy farming. Azores also produce wine and vegetables. In smaller islands, the meat production was more important than milk as is the case for Santa Maria and Corvo, but other agricultural productions were residual. In consequence, it is no surprise that milk production was the major product in value over 70.4% in relation to all farm production, in 1997, which has been improving to 76.2% in 1998 and 79.3% in 1999 (Marote and Silva 2011).

Azorean dairy policy depends on Common Agricultural Policy (CAP) of the European Union, namely, the milk production quota (about 547 million tons in 2011), although the Regional Government of Azores can adjust some rules on the regional agricultural policy. Some examples are measures to combat plagues and diseases, support to specific cultures needed for industrial transformation as sugar beet for sugar production and subsidies for biological production.

Azorean farms are small; some Azorean statistics shows about 8 ha per farm, about the half of the European average dimension (15.8 in 2003). The production system is primarily based on grazing (about 95% of the area). There were about 15,107 farmers in Azores. They were mainly old, more than 55 years old and low educated, mainly with basic education (4 years). This characterisation of the Azorean farms was based on national agricultural institution data and previous works of the authors (Silva et al. 1996).

The current historical context is particularly complex as some major changes are likely to occur. This is the case for the increased prices of some food products in international markets and, locally, the end of milk quota system. The multiplying effect of agriculture in both a small economy and the Azorean society makes this kind of work of major interest not only to protect the income of farmers but also to keep the society in equilibrium on employment matters and reduce immigration cycles. In this context, decision makers need information and knowledge for deciding the best policies in promoting quality and best practices.

The most important cost in Azores bovine farms are concentrates (variable cost and annual amortisation, fix cost). For 2002 data, the concentrates have the biggest value (about 30%) in milk systems and the lowest value in beef farms. The annual depreciation have increased in the last years and reached about 20% of total costs.

Fertilisers and land rent follow the importance in total cost (about 10%). The fertilisers get importance in beef systems with 16.6% of total costs and balanced the less use of concentrates (7.9%). The conservation and repair of equipment and construction vary from 4.9 (milk system) to 8.4% (mix system). The use of oil and petrol has values between 5.3 and 10.8%. Because the discriminated costs have more importance in total structure costs, it is supposed that they will affect more the efficiency of farms and they must be considered as variables for defining efficiency or inefficiency in farms.

The subsidies were an important part of the profit of dairy farms, and in 2004, it was about 61.6% of all profits in average per farm. Azorean agricultural farms had five main kinds of support from EU and government:

- Support to limit a plus production
- Maintaining extensification production, lake protection, protection of the genetic variability, etc.
- Money for more ecological production (the agri-environment measures)
- Support investment for planting trees in previously cultivated areas, maintaining tree culture, and support for less production as a result of planting trees
- Early retirement scheme

We investigate all 30 animal farms from Terceira Island of Azores. The initial list of potential variables is large. Any resource used by an Azorean dairy farm is treated as an input variable. The output variables come from the performance and activity measures that result when a farm converts resources to produce products. Following (Boussofiane et al. 1991) environmental variables which add resources are treated as inputs in our DEA models whereas those that require resources are treated as outputs. Applying DEA procedure, we focus on the choice of data variables in addition to the methodology of DEA.

The names of all input variables used in analysis are the following: *EquipmentRepair*, *Oil*, *Lubricant*, *EquipmentAmortization*, *AnimalConcentrate*, *VeterinaryAndMedicine*, *OtherAnimalCosts*, *PlantsSeeds*, *Fertilizers*, *Herbicides*, *LandRent*, *Insurance*, *MilkSubsidy*, *MaizeSubsidy*, *SubsidyPOSEIMA*, *AreaDimension* and *DairyCows*. The names of output variables are *Milk* and *Cattle*. After the outlier detection, one outlier identified in Terceira data was the result of a recording error and was corrected.

Canonical correlation analysis aims at highlighting correlations between input and output data sets. Two preliminary steps calculate the sample correlation coefficients and visualise the correlation matrixes. The correlation matrixes are visualised in Fig. 10.1.

Figure 10.1 highlights a significant correlation between *Milk* and *AnimalConcentrate* and nearly null correlation between *Milk* and *Lubricant*, *Milk* and *EquipmentAmortization* and *Milk* and *Insurance*.

In practice, the number of DMUs should be greater than the total amount of variables in both input and output sets. Any resource used by an Azorean dairy farm is treated as an input variable and because of it the list of variables that provide an accurate description of the milk and meat production process is large.

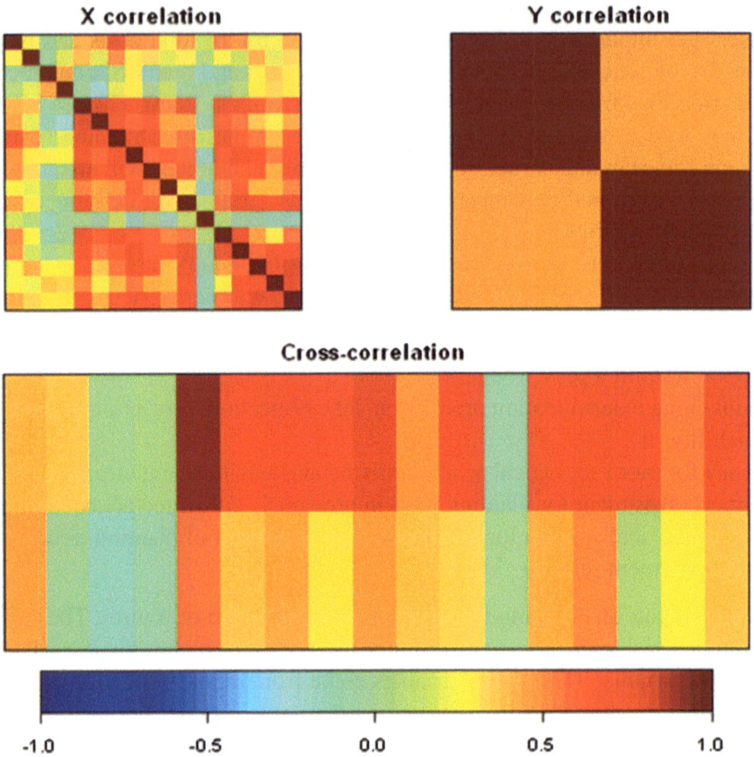

Fig. 10.1 Visualisation of sample correlation coefficients

This example is focused on measuring efficiency when the number of DMUs is few and the number of explanatory variables needed to compute the measure of efficiency is too large. We approach this problem from a statistical standpoint through both variable selection and variable aggregation approaches.

The results from CCA were already presented in the previous chapter of this book (Noncheva et al. 2012). From these results, we can conclude that both canonical variates were predominantly associated with the following original inputs: *AnimalConcentrate*, very big value almost a unit (−0.96); *Fertilizers* (−0.82); and with the original output variable *Milk*. These three variables are very strongly correlated, meaning that they contain the strongest dependence relation; they should be the major discriminant factors between farms. This way we selected the following two input variables *AnimalConcentrate* and *Fertilizers* and one output variable *Milk*. Note that, because the outputs have negative relations with the variates, negative values of the inputs should be directly related with the inputs and positive values are inversely related. There is very little negative correlation in the inputs and all are very weak, meaning that almost all the inputs contribute positively to the outputs.

The subsidies are very important in the farm income's output, especially maize and milk subsidies, respectively, −0.80 and −0.79. This result is also corroborated by statistical data as the Azorean SREA (2007b), where we can find that about 25% of the incomes of farmers were subsidies; this is a very big slice of the income. Also, if the subsidies were eliminated, the farm profit per month would be less than 400€ (inferior to minimum salary in Portugal). This shows the great importance of the subsidies in the sustainability of Azorean farms.

The number of dairy cows is more important than the farm dimension, respectively, −0.81 and −0.56, as it was expected because the number of cows should be more directly related with the milk production than the number of hectares, in spite of the effort towards extensive explorations.

In the previous chapter, "Azorean agriculture efficiency by PAR", we explore the concept of using CCA for aggregation of inputs and outputs. Here we use the same data to variable selection. Very similar results can be obtained.

10.6 Final Remarks

DEA models are used by PAR methodology to measure efficiency in production of Azoreans farms. DEA models are useful in situations in which multiple outputs are produced from a vector of inputs and no reliable price information exists that would allow estimation of stochastic frontier cost functions (Lovell 1993).

The "Productivity Analysis with R" (PAR) framework establishes a user-friendly data envelopment analysis environment with special emphasis on variable selection and aggregation and summarisation and interpretation of the results. The starting point is the following R packages: DEA (Diaz-Martinez and Fernandez-Menendez 2008) and FEAR (Wilson 2005). The DEA package performs some models of data envelopment analysis presented in Cooper et al. (2007). FEAR is a software package for computing non-parametric efficiency estimates and testing hypotheses in frontier models. FEAR implements the bootstrap methods described in Simar and Wilson (2008).

PAR is a software framework using a portfolio of models for efficiency estimation and providing also results explanation functionality. PAR framework has been developed to distinguish between efficient and inefficient observations and to explicitly advise the producers about possibilities for production optimisation. PAR framework offers several R functions for a reasonable interpretation of the data analysis results and text presentation of the obtained information. The output of an efficiency study with PAR software is self-explanatory.

It was applied, the PAR framework, to estimate the efficiency of the agricultural system in Azores (Noncheva et al. 2009). It is possible to rank observations (Azorean farms) in terms of their dissimilarity to other observations in the data (other Azorean farms). This makes PAR appropriate to support public policies in agriculture sector in Azores.

It is offered as a formal procedure for our intuitively sound approach to variable selection. This approach looks at the changes in the DMU's efficiencies when variables are added and removed from the DEA models in order to determine whether these changes are statistically significant. The procedure for variable selection has been developed by sequentially applying statistical and DEA techniques. This procedure is intended to produce DEA models that include only those variables that contribute to the closer input/output relations and have largest impact on the DEA results.

While this formal procedure can inform for the effect of adding and removing variables in a DEA study, the determination of the "best" model to represent any given situation must rely on managerial judgement and knowledge of the operations of the actual situation being represented.

It starts by selecting an initial model, involving all input and output variables. Next, the efficiency estimates for the initial model were compared to those for a new model in which some variables were subtracted. Efficiencies are calculated for each DMU under both the initial and reduced model. A statistical test was performed to determine whether the subtracting of some variables would significantly decrease the efficiency estimates. This procedure can be repeated until we receive a parsimonious model, using as many variables as needed but as few as possible.

PAR is used with both real and simulated data in order to find out a compromise between environment, agriculture and tourism and to investigate the potential impact of agricultural tourism on the farms' efficiency.

Acknowledgments This work has been partially supported by Direcção Regional da Ciência e Tecnologia of Azores Government through the project M.2.1.2/I/009/2008.

References

Alpert M, Peterson R (1972) On the interpretation of canonical analysis. J Mark Res 30:29–50

Banker R (1996) Hypothesis tests using data envelopment analysis. J Product Anal 7(2–3):139–159

Banker R, Charnes R, Cooper W (1984a) Some models for estimating technical and scale inefficiencies in data envelopment analysis. Manag Sci 30:1078–1092

Banker R, Charnes R, Cooper WW (1984b) Equivalence and interpretation of alternative methods for determining returns to scales in data envelopment analysis. Eur J Oper Res 89:473–481

Barcikowski R, Stevens J (1975) A Monte Carlo study of the stability of canonical correlations, canonical weights, and canonical variate-variable correlations. Multivar Behav Res 10:353–364

Bauernfeind U, Mitsche N (2008) The application of the data envelopment analysis for tourism website evaluation. Inf Technol Tour 10(13):245–257

Berger A, Humphrey D (1997) Efficiency of financial institutions: international survey and directions for future research. Eur J Oper Res 98:175–212

Boussofiane A, Dyson RG, Thanassoulis E (1991) Applied data envelopment analysis. Eur J Oper Res 52:1–15

Brockett PL, Golany B (1996) Using rank statistics for determining programmatic efficiency differences in data envelopment analysis. Manag Sci 42(3):466–472

Charnes A, Cooper WW (1985) Foundations of data envelopment analysis for Pareto-Koopmans efficient empirical production functions. J Econ 30(1/2):91–107

Charnes A, Cooper W, Rhodes E (1978) Measuring the efficiency of decision-making units. Eur J Oper Res 2:429–444
Chiang WE, Tsai M-H, Wang LS-M (2004) A DEA evaluation of Taipei hotels. Annal Tour Resear 31(3):712–715
Chilingerian J (1995) Evaluating physician efficiency in hospitals: a multivariate analysis of best practices. Eur J Oper Res 80:548–574
Cloutier L, Rowley R (1993) Relative technical efficiency: data envelopment analysis and Quebec's dairy farms. Can J Agric Econ 41:169–176
Cooper W, Seiford L, Tone K (2007) Data envelopment analysis: a comprehensive text with models, applications, references and DEA-solver software, 2nd edn. Springer, New York
Diaz-Martinez Z, Fernandez-Menendez J (2008) The DEA package, Version 0.1–2, Retrived from http://cran.r-project.org/web/packages/DEA/DEA.pdf
Färe R, Grosskopf S, Lovell C (1988) Scale elasticity and scale efficiency. J Inst Theor Econ 144:721–729
Fraser I, Cordina D (1999) An application of data envelopment analysis to irrigated dairy farms in northern Victoria, Australia. Agric Syst 59:267–282
Gifi A (1990) Nonlinear multivariate analysis. Wiley, Chichester
Gimenez-Garcia VM, Martínez-Parra JL, Frank P (2007) Improving resource utilization in multi-unit networked organizations: the case of a Spanish restaurant chain. Tour Manag 28:262–270
Golany B, Roll Y (1989) An application procedure for DEA. Technion-Israel Institute of Technology, Israel
Hallam D, Machado F (1996) Efficiency analysis with panel data – a study of Portuguese dairy farms. Eur Rev Agric Resou Econ 23(1):79–93
Hughes A, Yaisawarng S (2004) Sensitivity and dimensionality tests of DEA efficiency scores. Eur J Oper Res 154:419–422
Jaforullah M, Whiteman J (1999) Scale efficiency in the New Zealand dairy industry: a non-parametric approach. Aust J Agric Resour Econ 43(4):523–541
Jenkins L, Anderson M (2003) A multivariate statistical approach to reducing the number of variables in data envelopment analysis. Eur J Oper Res 147:51–61
Levine M (1977) Canonical analysis and factor comparison, vol 6, Quantitative applications in the social sciences series. Sage Publications, Thousand Oaks
Lewin A, Morey R, Cook T (1982) Evaluating the administrative efficiency of courts. Omega 10(4):401–411
Lovell C (1993) Production frontier and productive efficiency. In: Fried HO, Lovell CAK, Schmidt SS (eds) The measurement of productive efficiency-techniques and applications. Oxford University Press, Oxford, pp 3–67
Marianna S, David A, Peter J, Andrew L (2004) ICT paradox lost? A stepwise DEA methodology to evaluate technology investments in tourism settings. J Trav Resear 43(2):180–192
Marote E, Silva E (2002) Análise Dinâmica da Eficiência das Explorações Leiteiras da Ilha Terceira. XII Congresso de Zootecnia. November
Marote E, Silva E (2011) Importância dos Subsídios na Eficiência das Explorações Leiteiras da Terceira. Revista de Ciências Agrárias, pp 161–170
Noncheva V, Mendes A, Silva E (2009) An approach to variable aggregation in efficiency analysis. In: Classification, forecasting, data mining, international book series information science & computing. Suppl Int J Inf Technol Knowl 3(8):97–104
Noncheva V, Mendes A, Silva E (2012) Azorean agriculture efficiency by PAR. In: Mendes A, Silva E, Santos J (eds) Efficiency measures in the agricultural sector, with applications. Springer, Dordrecht, pp 53–72
Norman M, Stoker B (1991) Data envelopment analysis: the assessment of performance. Wiley, Chichester
Nunamaker T (1985) Using data envelopment analysis to measure the efficiency of non-profit organizations: a critical evaluation. Manag Decis Econ 6(1):50–58

OECD – Organisation for Economic Co-Operation and Development (2001) Environmental indicators for agriculture methods and results. Executive summary. Retrived from http://www.oecd.org/greengrowth/sustainableagriculture/1916629.pdf.
Pastor JT, Ruiz I-S (2002) A statistical test for nested radial DEA models. Oper Res 50(4):728–735
Rodriguez M, Gómez E, Lorente J (2004) Rural multifunctionality in Europe. The concepts and policies. 90th AEEA Seminar
Salinas-Jimenez J, Smith P (1996) Data envelopment analysis applied to quality in primary health care. Ann Oper Res 67:141–161
Sigala M, Airey D, Jones P, Lockwood A (2004) ICT paradox lost? A stepwise DEA methodology to evaluate technology investments in tourism settings. J Travel Res 43:180–192
Silva E, Santos C (2007) Eficiência nos Sistemas de Produção Pecuária nos Açores. APDEA Congress, Vila Real
Silva E, Arzubi A, Berbel J (1996) An application of data envelopment analysis (DEA) in Azores dairy farms. New Medit 3:39–43
Silva E, Arzubi A, Berbel J (2004) An application of data envelopment analysis (DEA) in Azores dairy farms. New Medit 3:39–43
Simar L (1996) Aspects of statistical analyses in DEA-type frontier models. J Product Anal 7:177–186
Simar L, Wilson P (2008) Statistical interference in nonparametric frontier models: recent developments and perspectives. In: Fried H, Lovell CAK, Schmidt S (eds) The measurement of productive efficiency and productivity change. Oxford University Press, New York
SREA (2007a) Estudo sobre os Turistas que visitam os Açores. 2005 – 2006. Região Autónoma dos Açores/ed. Serviço Regional de Estatística dos Açores
SREA (2007b) Anuário Estatístico dos Açores, 2007. Região Autónoma dos Açores/ed. Serviço Regional de Estatística dos Açores
Stevens J (1986) Applied multivariate statistics for the social sciences. Erlbaum, Hillsdale
Suhariyanto K (1999) Productivity growth efficiency and technical changes in Asian agriculture: a Malmquist index analysis. PhD thesis, University of Reading
Sungsoo P (2007) DEA application for the tourist satisfaction management. Tour Anal 12:201–211
Valdmanis V (1992) Sensitivity analysis for DEA model: an empirical example using public vs. NEP hospitals. J Public Econ 48:185–205
Venâncio F, Silva E (2004) A Eficiência de Exploração Agro-pecuárias dos Açores: uma abordagem paramétrica. XIV Jornadas Luso Espanholas de Gestão Científica
Wilson PW (2005) FEAR 1.0: a software package for frontier efficiency analysis with R, Retrieved from http://business.clemson.edu/Economic/faculty/wilson/courses/bcn/papers/fear.pdf

Chapter 11
The Importance of Subsidies in Azorean Dairy Farms' Efficiency

Emiliana Silva and Eusébio Marote

Abstract The purpose of this chapter is to analyze the importance of subsidies in the Azorean dairy farms efficiency from 1997 to 1999. The technical and economic variables of 82 dairy farms of the FADN (Farm Accountancy Data Network) were analyzed over the period of 3 years. The DEA (data envelopment analysis) was the approach used to calculate the efficiency. The software used was the DEAP (Data Envelopment Analysis Program).

The results show that the subsidies were not so important in the dairy farms' efficiency within 3 years. The technical efficiency variable and constant returns do not present great differences between model I (with subsidies as the output) and model II (without subsidies as the output). The number of efficient dairy farms was quite different, and the decreasing subsidies seem be compensated by the dairy production increase.

Keywords Subsidies • Efficiency • DEA • Milk • Azores

11.1 Introduction

The efficiency of FADN dairy farms was calculated using the nonparametric methodology, DEA, which allows the use of multiple outputs. In this research, two outputs were considered, the milk production and the subsidies. The software used was DEAP (Data Envelopment Analysis Program) developed by Coelli (1996).

In the first step, a summary of DEA approach is presented; in the second step, some subsidies that the Azorean farmers received from the European Union

---

E. Silva (✉) • E. Marote
CEEApIA, Departamento de Ciências Agrárias, Universidade dos Açores, Rua Capitão João de Ávila, 9700-042 Angra do Heroísmo, Açores, Portugal
e-mail: emiliana@uac.pt; pantera04@hotmail.com

A.B. Mendes et al. (eds.), *Efficiency Measures in the Agricultural Sector: With Applications*, DOI 10.1007/978-94-007-5739-4_11,

are presented. After, some characteristics of the Azorean FADN dairy farms will be presented. Finally, the DEA is applied, and the main results are shown and discussed.

11.1.1 The Common Agricultural Policy (CAP): Subsidies of Dairy Farms in the Azores

In the last years, the Azorean policy support was based in the decision no. 91/315/CE, 26th of June 1991, with POSEIMA – a specific actions program made for the isolated and insular nature of the Azores and Madeira – regulated by the European rules (Rule 1453/2001 and 1600/92). Another allowance is a compensation payment (Azorean administrative rules nos. 17/2001 and 37/2002) and finally arable crop support (Rule 1251/99). There was, also, some support for fuel and fertilizers logistic (Azorean administrative rules 7/2003) (Ministério da Agricultura 2003).

The POSEIMA is stipulated in the REG. (CE) no. 1453/01, which is specific for Azores and Madeira (JO L 198 of 21.07.2001).

The compensated payment is stipulated in REG (CE) no. 1257/99 of 17.05 – FEOGA-Rural Development (JO L 160 of 26.06.1999 and updated by JO L 302, of 01.12.2000).

Nowadays, all dairy farm support is integrated in the PRORURAL-Rural Development Plan presented from 2007 to 2013 to the European Union and is supported by the European Agricultural Fund for Rural Development (EAFRD).

11.1.2 Efficiency Approach: The Data Envelopment Analysis

The data envelopment analyses were developed by Charnes et al. (1978), and consequent developments of this methodology can be consulted in Damas and Romero (1997) and in Silva et al. (2002). The DEA model is usually presented in the following way:

$$\text{Max } E_j = \sum_{1}^{n} U_i Y_{ij}$$

s.a.

$$\sum_{1}^{m} V_i X_{ij} = \text{Constant}$$

$$\sum_{1}^{m} U_i Y_{ij} - \sum_{1}^{n} V_i X_{ij} \leq 0, \quad \forall_j$$

$$U_i \geq 0; \quad V_i \geq 0$$

E_j is the relative efficiency of the decision-making unit j, Y_{ij} is the level of output i used by decision-making unit j, X_{ij} is the level of input i used by decision-making unit j, and U_i and V_i are the nonnegative variable weights associated to the solution of decision-making unit j, of output and inputs, respectively. In this model, the decision-making unit is efficient relatively to other firms when its value is equal to unit (1).

The efficiency concept was initially introduced by Farrell (1957). He had decomposed the efficiency into two components: (1) technical efficiency that measures the maximum equiproportion reduction in all inputs that still allows continued production of given outputs and (2) allocative efficiency that reflects the ability of a firm to use the inputs in optimal proportions, given their respective price. Both these concepts form the concept of economics efficiency (Coelli 1995). The DEA estimates the efficiency using the mathematical programming (nonparametric), but it is possible to calculate the efficiency by recurring to an alternative econometric model (parametric). In the last years, various researchers chose the different approaches separately or a comparison between them. For instance, in the parametric model researches, the works of Brummer and Loy (2000), Reinhard and Thijssen (2000), Alvarez and González (1999), and Hallam and Machado (1996) are referred.

The nonparametric approaches were chosen by a lot of researchers because they have some advantages relatively to the parametric models. For instance, DEA (nonparametric) approach presents a great flexibility, has no implicit function form (less limitative technologies of reference), and is possible to adjust to a context of multiple outputs and inputs and the absence of prices.

Although, the DEA approach presents also some disadvantages. The conceptual difficulty to separate the noise from the error term is mainly referred. DEA was used in various researches in dairy farms, for instance, Marote and Silva (2002), Silva et al. (2002), Arzubi and Berbel (2002), Silva (2001), Jaforullah and Whiteman (1999), Fraser and Cordina (1999), Cloutier and Rowley (1993), and Gonzalez Fidalgo et al. (1996).

11.2 Materials and Methods

In the first place, 82 of Azorean FADN farms from 1997 to 1999 were characterized by the farms' income and costs composition.

In the cost composition, the main costs were with feeding (concentrated) which corresponded to about one third of the total cost. It was about 29.5% in 1997,

30.1% in 1998, and 30.2% in 1999. The machinery depreciation was the main fixed cost of the farms and the second most important cost of dairy farms' total costs. The machinery depreciation was about 12.5, 12.9, and 13.5% in 1997, 1998, and 1999, respectively. The third most relevant cost was fertilizers, about 11.7, 11.8, and 11.1% in 3 years (1997, 1998, and 1999). The rent is another important fixed cost with value above 10%. The other costs were a value less than 6% of total costs.

In the income composition, it was seen that the milk production was the main product, and it was 70.4% in 1997, increasing (76.2%) its importance in 1998, and in 1998, it was about 79.3%.

The subsidies, in opposite, decrease its importance in the total income, from 19.3% in 1997 to 11.8% in 1999. In 1998, the importance of subsidies in the total income was about 15.9%. The beef production was less than 10% in 3 years of the analysis and was usually complementary to the milk production.

The main components of income in the Azorean dairy farms, in 1997, 1998, and 1999, were the milk production and the subsidies. The beef production was a residual income and a subproduct of dairy farms. Along this period, the subsidies have decreased, and the milk production has increased, compensating the farmers' income. Consequently, the two outputs considered in the DEA approach were the milk production (physical unit) and subsidies (monetary unit).

The inputs integrated in DEA approach were consequent of the structure of total cost. Then, the costs selected were feed (concentrate), fertilizers, other specific animal cost (medicine, veterinary), fuel and lubricants, machinery and building repair, machinery depreciation, rent paid, and the agricultural area (hectares) and the number of animals in the dairy farms.

Two models were considered. In model I, two outputs and nine inputs were analyzed. The outputs were milk production (physical units) and the amount of subsidies (monetary units).

The inputs were agricultural area (physical unit) and number of animals (physical unit), and all other inputs were in monetary units: concentrate, fertilizers, other specific animal cost, fuel and lubricants, machinery and building repairs, machinery depreciation, and rent paid. In model II, there was only one output, the milk production, and the same nine inputs. The difference is the absence of the subsidies in model II.

11.3 Results and Discussion

11.3.1 Model I: Two Outputs and Nine Inputs

The average technical efficiency, a variable returns to scale (VRS), is high and very homogeneous (the standard deviation is less than 0.1). Its higher average value is about 0.960 in 1999, as seen in the next table (Table 11.1).

Table 11.1 The average technical efficiency of a variable returns to scale (VRS) in the dairy farms from 1997 to 1999

Efficiency VRS	1997	1998	1999
Average	0.957	0.951	0.960
Standard deviation	0.076	0.085	0.095
Minimum	0.637	0.620	0.504
No. of efficient firms	52	51	58
No. of efficient firms in 3 years	34		

Fig. 11.1 Efficiency intervals of a VRS in dairy farms from 1997 to1999

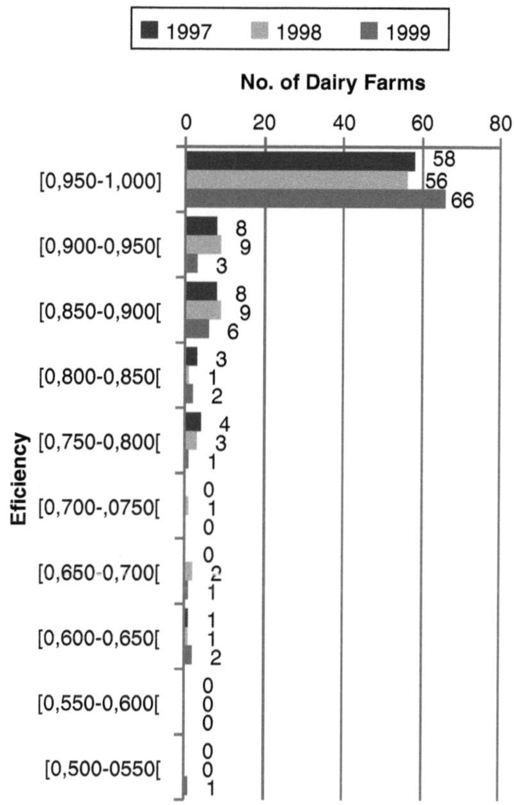

The average of technical efficiency of a VRS is higher (0.950) in 3 years, although it is possible that some farms increased the efficiency of its VRS, as the minimum value confirms it (from 0.504 to 0.637). That means that these farms could produce the same using fewer inputs. The number of efficient farms goes from 62% (51/82) to 71% (58/82), and there were 34 farms that were simultaneously efficient in the three periods; they are about 41% (34/82) of dairy farms.

In Fig. 11.1, the mode for all periods is quite similar, and the majority of farms lie in the interval [0.950–1.000]. There are no efficient farms above the interval [0.500–0.550].

Table 11.2 Model I: The average technical efficiency of a constant returns to scale (CRS) in the dairy farms from 1997 to 1999

Efficiency CRS	1997	1998	1999
Average	0.927	0.929	0.930
Standard deviation	0.091	0.100	0.106
Minimum	0.629	0.577	0.504
No of efficient firms	39	42	39
No. of efficient firms in 3 years	21		

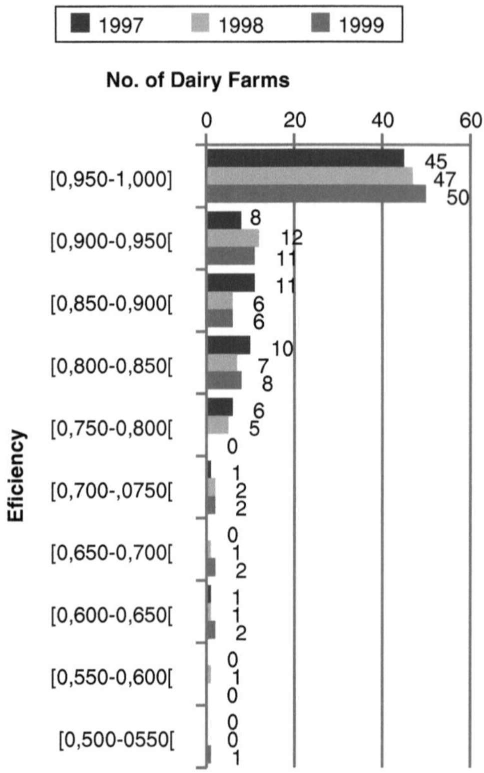

Fig. 11.2 Model I: Efficiency intervals of a CRS in dairy farms from 1997 to 1999

There are fewer farms above the interval [0.850–0.900] (Fig. 11.1). The most part of farms has a high efficiency, and only a few farms are not efficient.

The efficiency of a CRS is high along the three periods, and it is about 0.93, and it is less than the efficiency of a VRS (Table 11.2). Although some farms could have a better performance as it is seen by the minimum value of efficiency (from 0.504 to 0.629), that means it is possible to reach the same output using less inputs. The efficient farms are 48% (39/82) and 51% (42/82). There were about 21 farms operating efficiently within 3 years, about 26% (21/82) of the total farms.

The efficiency mode value is similar, and it corresponds to the interval [0.950–1.000] (Fig. 11.2). There were no efficiency values above the interval [0.500–0.550] and only a few farms above the efficiency interval [0.800–0.850].

Table 11.3 Model II: The average technical efficiency of a constant returns to scale (CRS) in the dairy farms from 1997 to 1999

Efficiency CRS	1997	1998	1999
Average	0.905	0.922	0.916
Standard deviation	0.114	0.104	0.125
Minimum	0.415	0.577	0.409
No. of efficient firms	31	38	32
No. of efficient firms in 3 years	16		

Table 11.4 The average technical efficiency of a variable returns to scale (VRS) in the dairy farms from 1997 to 1999

Efficiency VRS	1997	1998	1999
Average	0.943	0.946	0.953
Standard deviation	0.091	0.091	0.107
Minimum	0.611	0.592	0.461
No. of efficient firms	45	49	55
No. of efficient firms in 3 years	28		

11.3.2 Model II: One Output and Nine Inputs

In model II, the efficiency of a CRS is still high but lower than in model I. In model II, the efficiency value varies from 0.905 to 0.922 in the years 1997, 1998, and decrease in the year 1999 (Table 11.3).

And the same as in model I, some farms could improve its efficiency and produce the same with less use of inputs.

The efficient farms' range goes from 38% (31/82) to 46% (38/82), but it still is lower than in model I (Table 11.3). There have been 16 efficient farms within 3 years, which corresponds to about 19.5% (16/82) of the total of dairy farms (Table 11.4).

Almost all Azorean dairy farms lie in the interval [0.950–1.000] (Fig. 11.3) as observed in model I. The efficiency mode is higher in model I. There were few farms above the interval [0.750–8.00], and there were no dairy farms above the interval [0.400–0.450].

The efficiency of the VRS was high along the period 1997–1999, and its value is about 0.943 and 0.953. Despite this, there were some farms that could improve their efficiency (the minimum value of efficiency ranges from 0.461 to 0.611). The number of efficient farms goes from 55% (45/82) to 67% (55/82). There were 28 farms which always have been efficient within 3 years, but this value is superior in model I.

As referred in the bibliography, the average efficiency value of the VRS is higher than the CRS (Fig. 11.4).

A comparison between model I (with subsidies) and model II (without subsidies) was made but with the existing variables applied. The value of efficiency is similar and goes around 92%. The number of efficient farms is similar too in both models.

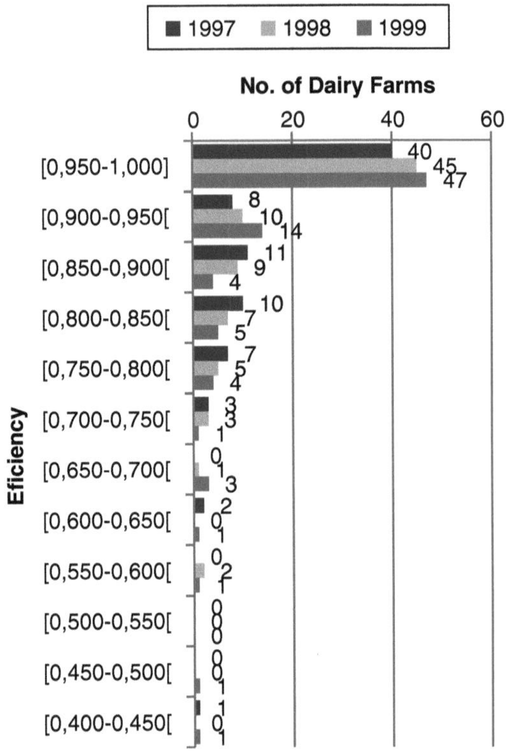

Fig. 11.3 Model II: Efficiency intervals of a CRS in dairy farms from 1997 to 1999

Analyzing both models, it is concluded that in this case, the subsidies do not improve the efficiency value in the Azorean dairy farms.

11.4 Conclusions

Using DEA, the average technical efficiency reached a high value in the period of 1997–1999.

Although the subsidies are a great contribution in the farmer's income, its influence in the efficiency is not very relevant because of the number of efficient farms, and the value of efficiency does not experience great changes.

Some dairy farms could improve their efficiency if they rationally used their resources in the feeding and machinery equipments' costs.

To increase the efficiency, it is required to know individually each farm and to identify the resources that cause inefficiency. In this case, having a controlled feeding and a better adjustment of the equipment to the dimension of the farms is suggested.

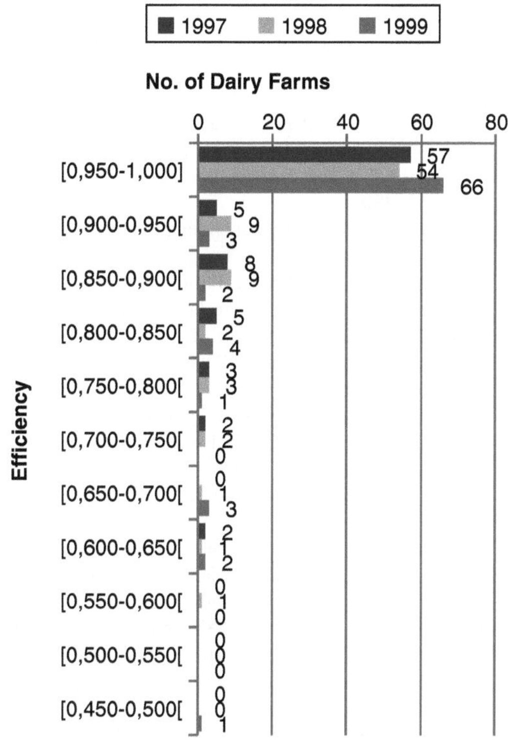

Fig. 11.4 Efficiency intervals of a VRS in dairy farms from 1997 to 1999

The comparison with other researches shows that in the Azores dairy farms, efficiency is greater, although this can be explained by the use of a high number of inputs (9). Suhariyanto (1999) suggests that a greater use of variables could improve the value of efficiency.

References

Alvarez AM, González E (1999) Using cross-section data to adjust technical efficiency indexes estimated with panel data. Am J Agric Econ 81:894–901

Arzubi A, Berbel J (2002) Determinación de la Eficiencia usando DEA en Explotaciones Lecheras en Argentina, IV Congreso de la Asociación Española de Economía Agraria, "Economia Agraria y Recursos Naturales: Nuevos Enfoques y Perspectivas", Pamplona, España

Brummer B, Loy J (2000) The technical efficiency impact of farm credit programs – a case study of Northern Germany. J Agric Econ 51(3):405–418

Charnes A, Cooper WW, Rhodes E (1978) Measurement the efficiency of decision making units. Eur J Oper Res 2:429–444

Cloutier L, Rowley R (1993) Relative technical efficiency: data envelopment analysis and Quebec's dairy farms. Can J Agric Econ 41:169–176

Coelli TJ (1995) Recent developments in frontier modelling and efficiency measurement. Aust J Agric Econ 39(3):219–245

Coelli T (1996) A guide to DEAP version 2.1.: a data envelopment analysis computer program. Centre for Efficiency and Productivity Analysis, Department of Econometrics, University of New England, Armidale

Damas Rico E, Romero C (1997) Análisis no paramétrico de la eficiencia relativa de las Almazaras Cooperativas en la Provincia de Jaén Durante el Período 1975–1993. Revista Española de EconomíaAgraria 180(2):263–279

Farrell M (1957) The measurement of productive efficiency. J R Stat Soc (Series A) 120(part III):253–290

Fraser I, Cordina D (1999) An application of data envelopment analysis to irrigated dairy farms in Northern Victoria, Australia. Agric Syst 59:267–282

Gonzalez Fidalgo E, Alvarrez Pinilla A, Arias Sampedro C (1996) Análisis no paramétrico de eficiencia en explotaciones lecheras. InvestigaciónAgraria 11(1):173–190

Hallam D, Machado F (1996) Efficiency analysis with panel data- a study of Portuguese dairy farms. Eur Rev Agric Econ 23(1):79–93

Jaforullah M, Whiteman J (1999) Scale efficiency in the New Zealand dairy industry: a non-parametric approach. Austr J Agric Resour Econ 43(4):523–541

Marote E, Silva E (2002) Análise Dinâmica da Eficiência das Explorações Leiteiras da Ilha Terceira, Actas do XII Congresso de Zootecnia

Ministério da Agricultura (2003) Retrieved from http://www.inga.min-agricultura.pt/ajudas

Reinhard S, Thijssen GJ (2000) Nitrogen efficiency of Dutch dairy farms: a shadow cost system approach. Eur Rev Agric Econ 27(2):167–186

Silva E (2001) Análisis de la Eficiencia Económica de las Explotaciones Agroganaderas de las Azores (Portugal). Tesis Doctoral, Universidad de Córdoba

Silva E, Berbel J, Arzubi A (2002) Análisis no Paramétrico de eficiencia en las Explotaciones Lecheras de las Azores, a partir de Datos RICA-A, IV Congreso de la Asociación Española de Economía Agraria, "Economía Agraria y Recursos Naturales: Nuevos Enfoques y Perspectivas", Pamplona, España

Suhariyanto (1999) Productivity growth efficiency and technical changes in Asian agriculture: a Malmquist index analysis. PhD thesis, University of Reading

Chapter 12
Multi-output Technical Efficiency in the Olive Oil Industry and Its Relation to the Form of Business Organisation

Rafaela Dios-Palomares, José M. Martínez-Paz, and Angel Prieto

Abstract This work studies the level of technical efficiency in the Andalusian oil industry from a multi-output, non-parametric approach by conducting the data envelopment analysis (DEA) methodology with non-radial distance functions, as well as implementing environmental and non-discretionary variables. The production frontier includes three outputs: quantity and quality of oil production, the outputs to be maximised, and one output to be minimised, the environmental impact of the production process. The inputs are the following: grinded olive, labour, and capital (both fixed and floating). The analysis is carried out by including non-discretionary variables from two points of view. It is considered that the business structure (cooperative or corporation) of the firm affects the frontier (technology). This variable is included through a specific three-stage method. The relation between efficiency and other non-discretionary variables is analysed by the estimation of a Tobit model. Having a sample of 88 oil-mill industries in Andalusia as the starting point, the indices for the two nonconventional outputs in this type of analysis are elaborated; quality is quantified by means of an aggregated index that gathers some aspects related to the separation of olives, critical points, and traceability. The environmental impact is assessed by another index that includes the effects produced on soil, water, air, and sound comfort. From the analysis of results, it can be underlined that, in spite of the fact that the levels of efficiency are high

R. Dios-Palomares (✉)
University of Córdoba, Campus de Rabanales, Córdoba, Spain
e-mail: rdios@uco.es

J.M. Martínez-Paz
University of Murcia, Campus Espinardo, Murcia, Spain
e-mail: jmpaz@um.es

A. Prieto
IRNASA. CSIC, Salamanca, Spain
e-mail: angel.prieto@irnasa.csic.es

A.B. Mendes et al. (eds.), *Efficiency Measures in the Agricultural Sector: With Applications*, DOI 10.1007/978-94-007-5739-4_12,

on average, some production adjustments to reduce inputs and the environmental impact of the process could be implemented. The influence of the business structure is significant, and results show that corporations are the most effective ones.

Keywords Olive oil • Cooperative • DEA • Technical efficiency • Hyperbolic distance • Environmental impact • Non-discretionary variables

12.1 Introduction

Olive oil is a product of particular importance within the Mediterranean agricultural food system, and more specifically in Spain, owing to two main reasons: first, because olive oil is an essential component of the so-called Mediterranean diet and, second, because Spain, and Andalusia in particular, is the world's main production area; over the last 10 years (2001–2010) around 39% of the olive oil produced worldwide is Spanish in origin, and over 79% of this oil comes from Andalusia (Rallo 2010).

The olive oil production industry, which is the object of this study, is the core of a production chain that starts in the olive sector, the producer of olives, and ends at the olive oil packaging and marketing sector. This production area is highly regulated by the EU Common Agricultural Policy (CAP) through the common organisation of the market in olive oil, pursuant to Commission Regulation (EC) 136/66, and its successive reforms in 1998 (Commission Regulation (EC) 1638/98), 2004 (Commission Regulation (EC) 865/2004), and 2007 (Commission Regulation (EC) 1234/2007). These regulations follow the trend put forward by the CAP reform in 1992. It tends towards the gradual reduction of direct support to production, which is the reason why the oil industry must necessarily increase its direct profitability. Thus, efficiency and productivity levels have to increase, and trade policies devoted to open new competitive markets have to be put into practice in order to compensate for the continuing decrease in support, which will take place in the near future (Mili 2009). The social, financial, and environmental importance of the olive sector in Spain, and especially in Andalusia, is a widely known and researched issue (Sánchez et al. 2011). Nearly 90% of all olive production (Rallo 2010) is used to produce oil, which means that the future of the olive sector is closely related to and conditioned by the future of its extraction industry. Therefore, the improvement of the industrial production processes has a direct impact on the enhancement of the agricultural sector that provides the raw material.

There is a highly relevant and differential feature in this industry, namely, the existence of two types of business structures: cooperatives, made up by the olive producers themselves, who associate to create the extraction industry of their own production, and corporations, which are not linked to any olive grove owners and in most cases are part of big food companies that buy the raw material directly from the market. Given their special ownership structure, cooperatives (Brada and Méndez 2009) introduce a distinguishing factor in the management system of their inputs and outputs that is different from that in traditional companies (Soboh et al. 2009). Some of these differences are the periods for payment to clients and providers, the profit-sharing mechanism, productive investments, etc. (Cook 1995). Some authors point out that it is to be expected that cooperatives have a lower level of technical efficiency than corporate companies (Salazar and Galve 2008), owing to the elevated management costs of the labour factor, among others (Bartlett et al. 1992; Schmitt 1993). Other works hold that the democratic decision-making system might restrict efficiency due to the heterogeneous and, sometimes, clashing interests of the owners (Jensen and Meckling 1979). At an empirical level, there are some studies that show this negative relation (Piesse et al. 1996; Ferrier and Porter 1991; Barreiro et al. 2009; Barnes 2006), but there are others that find a positive relation between cooperativism and technical efficiency (Hart and Moore 1990; Hofler and Payne 1993; Maietta and Sena 2008), as well as some other works that do not establish any determining relations (Bonin et al. 1993; Jones 2007; Alonso and García 2009).

The efficiency in production has been analysed very often in the field of technical efficiency through the production frontier function, as well as in the field of assignative and economic efficiency, considering the frontier of costs or profits as the base. Nowadays, the most used methodologies for the efficiency estimation through the frontier function are the following: the mathematic programming by the data envelopment analysis or DEA (Cooper et al. 2004) and the so-called econometric frontier (Battese 1992). The average efficiency level of the sample and the efficiency index of each company can be estimated by using both methods.

The study of the efficiency in the agricultural sector has a long-standing tradition, and this can be seen in several meta-analyses: Bravo-Ureta and Pinheiro (1993), analysing 39 cases; Abdourahmane et al. (2001), gathering 51 estimations of technical efficiency from 32 works; or Bravo-Ureta et al. (2007), relating 167 studies on technical efficiency at farm level. In Spain, a large number of empirical applications have been carried out since the late 1980s. Many of them are included in Alvarez's work (2001). Dealing with the Spanish olive production sector, the recent works by Amores and Contreras (2009) are also relevant, as they analyse the efficiency of olive groves in Andalusia. On the other hand, Lambarraa et al. (2007, 2009) study the technical efficiency and the productivity increase in the Spanish olive groves. At the international level, the most important recent works are those by Giannakas et al. (2000), Tzouvelekas et al. (2001), Karagiannis et al. (2003), and Karagiannis and Tzouvelekas (2009). All of them deal with technical efficiency of the olive sector in Greece. There are few analyses on the efficiency of the oil-mill industry. Some of them are the ones performed by Millán (1986), Damas and Romero (1997), and Dios-Palomares and Martínez Paz (2011).

The main goal of this research is to measure the technical efficiency of this sector, which is the previous step before assessing the improvement opportunities in the resource management of this industry. In order to achieve this goal, the indicators of the levels of environmental impact and quality control of the production process have been designed and constructed, since these two aspects are key factors for the future of the sector and to get a quality product that allows implementing differentiation and segmentation strategies of the market

(Gázquez and Sánchez 2009). On the other hand, these points are also crucial for the processes to be environmentally sustainable, as current administrative regulations state, because of the advantages of the competitive image, since consumers are even more aware of these issues (Mesias et al. 2011).

This work incorporates three types of nonconventional variables into the analysis, as well as the ones that take part in the production process. In the first place, the environmental impact is a variable included in the analysis as an output to be minimised, whereas quality is an output to be maximised. The estimation of the global efficiency is carried out by means of an envelopment analysis model with hyperbolic distance. Although this methodology (DEA environmental modelling) has been previously applied in other sectors by Ball et al. (2004) and Hernández-Sancho et al. (2000), among others, it has never been applied to the olive oil area.

In the second place, the effects of the form of business organisation (hereafter also referred to as FBO) are analysed taking into account that this variable has an impact in the frontier, meaning that cooperatives have a different production technology from that of trading companies. It is considered a non-discretionary variable for it to be included. Several works have tackled this issue from different points of view in order to solve the problem. Some of the most relevant ones are the following Banker and Morey (1986), Lozano-Vivas et al. (2002), Muñiz (2001), Fried et al. (1999, 2002), and Daraio and Simar (2005), for continuous variables. In our case, FBO is a categorical variable, and we applied a method that is wider than that by Charnes et al. (1981). Our three-stage process allows estimating the efficiency once the effect of this variable has been removed. The application of this system in the same sector is seen in Dios-Palomares et al. (2005). This research puts forward a new global approach where the analysis is conducted by taking both aspects into account at the same time. In this way, the three-stage method, thoroughly described in the next section, is implemented in order to correct the impact of FBO on efficiency by applying the environmental modelling with directional distances in each phase.

And last, this work also studies the impact that other non-discretionary, socioeconomic variables (like seniority, number of members, and so on) may have on efficiency. An econometric model is estimated to determine whether there are any relations between these variables and the resource management carried out by the company (efficiency). Given the special structure of the endogenous variable, the specification of a general linear model has been criticised in specialised literature (Simar and Wilson 2007). That is the reason why this work estimates a Tobit model. The conclusions of this relation allow suggesting several strategies to improve the efficiency levels of the sector.

After this introduction, this study goes on with the theoretical description of the methodology used to study efficiency. The third section deals with more specific methodological aspects such as the source of the data, the construction process of the quality and environment indicators, or the formulation of the efficiency model. Results are shown in the following section, and the final one includes a summary and the conclusions of this study. The bibliography is in the last section of the document.

12.2 Methodology

In this section, the theoretical foundations of the efficiency model applied in this work are introduced. This model is a DEA with a hyperbolic distance function and environmental, categorical, and non-discretionary variables. To get this, the efficiency measuring methodology is combined with environmental non-discretionary variables (Dios-Palomares et al. 2006), with non-radial distance functions (Prieto and Zofío 2004), which allows to simultaneously include both the minimisation of the undesirable output and the effect of different technologies in the efficiency index of each company.

12.2.1 Characterising the Environmental Production Possibility Set

The conventional measurement of efficiency deriving from the original partially oriented output or input DEA models is not well suited to assess performance in an environmental framework, where undesirable outputs are produced. The reason is the radial equiproportional expansion of both sets of outputs – desirable and undesirable – in a "business as usual" strategy.[1] Nevertheless, since the work by Färe et al. (1989), it is possible to make use of the so-called hyperbolic distance function that considers both outputs asymmetrically by expanding the desirable outputs while reducing the undesirable outputs, marketed or not (Baumgärtner (2004)), giving rise to what is now known as DEA environmental modelling.

The definition of the hyperbolic distance function allows characterising the technology in a joint-production setting that is based on physical thermodynamic laws and, therefore, accounting for the non-separability between both sets of desirable and undesirable outputs, that is, null jointness is a necessary assumption for the production technology (Faber et al. 1998).[2] DEA environmental modelling allows establishing control and management programmes of contaminants and pollutants based on best practice criteria. This methodology is well developed in Ball et al. (1994, 2004), Hernández-Sancho et al. (2000), Zaim and Taskin (2000), Färe and Grosskopf (2004), Zofío and Prieto (2001), Prieto and Zofío (2004), Zaim (2004), Zhou et al. (2007), and Picazo-Tadeo and Prior (2009).

The formulation of the undesirable output reduction and the desirable output expansion in a joint-production framework by way of programming techniques has been accomplished by means of two hypotheses. These characterise the

[1] Actually CCR and BCC models cannot be applied in this case. The direction to the frontier of desirable outputs must differ from that of undesirable outputs.

[2] "The concept of joint production captures essential physical aspects of production. To this end, we want to link the economic concept of joint production to the laws of thermodynamics, and in particular the Entropy Law", p. 132.

environmental production possibility set, defining a joint environmental technology (Färe et al. 2005) that can be implemented through its environmental DEA counterpart (Zhou et al. 2007). This formulation leads to a specific method of modelling the undesirable outputs that comes from the traditional DEA structure that deals with them as though they were inputs, as in Hailu and Veeman (2001), Seiford and Zhu (2002), and Färe and Grosskopf (2004).[3]

To present our DEA model and assess environmental performance in the olive oil-mill industry, we first introduce the required concepts and notation. Let us assume that there exists a production process that transforms a vector of inputs $x = (x_1, \ldots, x_N) \in R_+^N$ into a vector of outputs $y = (y_1, \ldots, y_M) \in R_+^M$, which can be partitioned into two desirable and undesirable output subvectors: $p = (p_1, \ldots, p_P) \in R_+^P$ and $q = \left(q_1, \ldots, q_Q\right) \in R_+^Q$, respectively.

The environmental production possibility set is defined as

$$T = \{(p, q, x) \colon x \text{ can produce } (p, q)\}, \tag{12.1}$$

and it is assumed that T satisfies basic regularity properties: compact for each $x \in R_+^N$ – bounded and closed or interior, positive production cannot be possible without consumption inputs (i.e. free production is excluded) and free disposability of inputs and outputs (i.e. it is possible that an increase in inputs cannot originate an increase in outputs); see Färe et al. (1985) for its axiomatic formulation.

The technology set T can be equivalently expressed in terms of the output correspondence:

$$x \to P(x) \subseteq T_+^M, \quad M = P + Q, \tag{12.2}$$

where $P(x)$ denotes all (p, q) output vectors that can be produced by using the x input vector.

Given the regularity conditions of T, allowing $P(x)$ to represent an environmental production possibility set requires the assumption that it is not possible to produce q without p (i.e. the null jointness assumption):

1. If T allows for the joint production of (p, q) from a given vector x, then

$$(p, q) \notin P(x), \ \ \forall q = 0, \ \ p \neq 0; \quad \text{or} \quad (p, q) = (0, 0) \in P(x). \tag{12.3}$$

2. If for a given vector x, it is not possible to reduce q without reducing p, then the reduction of q bears a cost on p (i.e. a technological opportunity cost). This assumption is formally introduced by considering that q is weakly disposable with regard to p:

[3]The proposal to treat undesirable outputs as inputs by DEA can lead to undesirable outputs without desirable outputs, (Zhu 2009), violating the null jointness assumption. This treatment is compiled by using appropriate mathematical programming techniques in the context of DEA models without outputs (Lovell and Pastor 1995).

$$\text{If } (p,q) \in P(x) \Rightarrow (\lambda p, \lambda q) \in P(x), \quad 0 < \lambda \leq 1. \tag{12.4}$$

When a specific industrial technology requires it, we can also allow for free disposability for p and x:

$$\text{If } p \in P(x), \quad p \geq p' \Rightarrow p' \in P(x), \quad \text{and} \tag{12.5}$$

$$\text{If}(p,q) \in P(x), \quad x' \geq x \Rightarrow (p,q) \in P(x'), \tag{12.6}$$

the output vector (p, q) produced by a smaller input vector also belongs to the production possibility set associated to a larger input vector.

For performance measurement purposes, we can define two alternative subsets in the environmental production possibility set, $P(x)$ – satisfying properties (12.3) and (12.4):

$$\text{Isoq } P(x) = \{(p,q): (p,q) \in P(x), \lambda > 1, (\lambda p, \lambda q) \notin P(x)\}, \tag{12.7}$$

it is not possible to increase (p, q) for a given x, but it is possible to increase (or to diminish) q for a given p, and

$$\text{Eff } P(x) = \left\{(p,q): (p,q) \in P(x), (p',q') \geq (p,q) \Rightarrow (p',q') \notin P(x)\right\} \tag{12.8}$$

From (12.7) and (12.8), it is possible to see that for any vector,

$$(p',q'), p' \geq p, q' \leq q \Rightarrow (p',q') \notin P(x).$$

Assumption (12.3) means that the projection towards the origin of any observed output vector (p, q) belongs to the production possibility set, and therefore, in the limit of that projection, the origin belongs to it, while (12.4) assumes that activity in the undesirable output axis is not possible.

12.2.2 The Hyperbolic Measure of Environmental Productive Performance

Once the environmental production possibility set $P(x)$ has been defined, our interest focuses on how to measure the distance separating any decision unit from the production frontier of $P(x)$. Färe et al. (1985) introduced the hyperbolic distance function (HD) in order to measure the possibility of expanding the desirable outputs while reducing the inputs at the same time. They called it hyperbolic because the asymmetric projection path towards the production frontier corresponds to that definition, constituting a generalisation of the partially radial distance functions. The seminal work by Färe et al. (1989) adapts this setting to the environmental case by replacing the undesirable output vector q for the input vector x, allowing for equiproportional undesirable output reduction and desirable output expansion. In this vein, the main advantage of the hyperbolic measurement framework is that it accommodates the relevant environmental information such as the change in the ratio p/q.

To empirically determine the efficient frontier and to calculate the efficiency scores reflecting the environmental behaviour of the firms, we assume that there exist observations $k = 1, \ldots, K$ – DMU's in the operations research literature. For a particular firm "o", $P(x)$ is defined as

$$P(x_o) = \left\{ (p, q): \sum_{k=1}^{K} z_k p_{kp} \geq p_{oq}, \sum_{k=1}^{K} z_k q_{kq} = q_{oq}, \sum_{k=1}^{K} z_k x_{kn} \leq x_{on}, z_k \in R_+^K \right\}, \tag{12.9}$$

where z_k are intensity variables for the different linear combinations, which therefore represent weights for each k in $P(x_o)$.

It is assumed that the degree of efficiency of every firm corresponds to the value of the hyperbolic distance function (HD):

$$\mathrm{DH}(p, q) = \max \left\{ \theta : \left(\theta p, \frac{q}{\theta} \right) \in P(x), \quad \theta \geq 1 \right\} \tag{12.10}$$

From (12.9) and (12.10), the environmental DEA efficiency score for each company k can be calculated by solving

$$\begin{aligned} &\mathrm{DH}(p_o, q_o, x_o) = \max_{z\theta} \theta \\ &Pz \geq \theta p_o \\ &Qz = \frac{q_o}{\theta} \\ &Xz \leq x_o \\ &z \geq 0 \end{aligned} \tag{12.11}$$

Where $\theta \geq 1$,

- $z = (z_1, z_2, \ldots, z_K)^t \in \Re^K$.
- $P = (p_1, p_2, \ldots, p_K) \in \Re^{P \times K}$ is a $P \times K$ matrix of desirable outputs (with $p_j \in \Re^P$ the data vector of the output values at DMU_j).
- $Q = (q_1, q_2, \ldots, q_K) \in \Re^{Q \times K}$ is a $Q \times K$ matrix of undesirable outputs (with $p_j \in \Re^Q$ the data vector of the output values at DMU_j).
- $X = (x_1, x_2, \ldots, x_K) \in \Re^{N \times K}$ is a $N \times K$ matrix of inputs (with $x_j \in \Re^N$ the data vector of the input values at DMU_j).

Equation (12.11) corresponds to a model specified under the assumption of constant returns to scale (CRS). Variable returns to scale[4] can be imposed by adding the following restriction:

$$\sum_{k=1}^{K} z_k = 1$$

Comparing the solutions to both models, we can analyse the impact of the productive scale in the level of efficiency.

These models do not distinguish between (12.7) and (12.8) and do not exclude the possibility of increasing (or reducing) q for a given p; desirable output and input slacks may exist, implying non-equiproportional reduction in p. This problem can be tackled by adopting a two mathematical programming approach (Ali and Seiford 1993). In fact, slacks can be calculated by solving the following second-stage problem:

$$\begin{aligned}
&\max_{z, S_p^+,\, S_q^-, S_x^-} S_p^+ + S_q^- + S_x^- \\
&Pz - S_p^+ = \theta^e p_o \\
&Qz + S_q^- = \frac{q_o}{\theta^e} \\
&Xz + S_x^- = x_o \\
&z \geq 0
\end{aligned} \tag{12.12}$$

where S_p^+, S_q^-, S_x^- stand for the outputs p, q, and input slacks and θ^e is the efficiency score obtained when solving (12.11).[5]

12.2.3 The Three-Stage Programme Method

Once the environmental production possibility set $P(x)$ and hyperbolic measure of environmental productive performance with respect to the frontier have been defined, our aim is to estimate the efficiency level regarding the undesirable outputs by taking non-discretionary variables into account as well. Our method is developed in three stages in order to isolate the effect of these variables.

---

[4]It is possible to go deeper in the concepts of returns to scale in Cooper et al. (2004), Chapter 2, p. 41.

[5]Expression (12.11) is not linear because of the second set of restrictions, but it is easily computable in non-linear programming. All these models were specified in MATLAB using the non-linear optimiser "fmincon" – find minimum of a constrained non-linear multivariate function.

12.2.3.1 Estimation of Environmental Efficiency Regarding a Categorical Non-discretionary Variable

To describe the applied method, we consider the case of a dichotomous, non-discretionary variable without loss of generality. The method is addressed to a group of K companies, where the variables to consider in the efficiency analysis are P desirable outputs (p_p), Q undesirable outputs (q_q), and N inputs (x_n), with a dichotomous, non-discretionary variable z with values z_h for $h = a, b$. Regarding this variable, the sample is divided into two subsamples, K_h in size, and their data matrices would be for each subsample h *(for* $h = a, b$*):*

- $P_h = \left(p_{1h}, p_{2h}, \ldots, p_{K_h h}\right) \in \Re^{P \times K_h}$ is a $P \times K_h$ matrix of desirable outputs (with $p_{kh} \in \Re^P$ the data vector of the output values at DMU$_{kh}$ belonging to the subsample where $z_h = h$).
- $Q_h = \left(q_{1h}, q_{2h}, \ldots, q_{K_h h}\right) \in \Re^{Q \times K_h}$ is a $Q \times K_h$ matrix of undesirable outputs (with $p_{kh} \in \Re^Q$ the data vector of the output values at DMU$_{kh}$ belonging to the subsample where $z_h = h$).
- $X_h = \left(x_{1h}, x_{2h}, \ldots, x_{K_h h}\right) \in \Re^{N \times K_h}$ is a $N \times K_h$ matrix of inputs (with $x_{kh} \in \Re^N$ the data vector of the input values at DMU$_{kh}$ belonging to the subsample where $z_h = h$).

Next the method is analysed by describing its three stages.

I. *Stage One*

The sample is divided into the two different subsamples for $h = a$ and $h = b$, and a frontier is estimated for each of them, applying the expressions (12.11) and (12.12), which allows achieving the HD function for the environmental DEA model with the aim of obtaining the intra-group efficiencies, which are named

$$\theta_{kh} \text{ for } kh = 1, \ldots, K_h, \text{ and } h = a, b \text{ and the slacks } S^+_{pkh}, S^-_{qkh} \text{ and } S^-_{xkh}.$$

From now on, we substitute the observed data values for their target values (projected onto the frontier), each one in its corresponding subsample. By doing this, we eliminate inefficiency that is relative to each unit in its group. New values for outputs and inputs are calculated for each value of h (a and b), according to the following correction:

$$P_h{}^* = \left(p^*_{1h}, p^*_{2h}, \ldots, p^*_{K_h h}\right) \in \Re^{P \times K_h} \text{ with } p^*_{kh} = p_{kh}\theta_{kh} + S^+_{pkh} \text{ and } k = 1 \ldots K_h$$
$$Q_h{}^* = \left(q^*_{1h}, q^*_{2h}, \ldots, q^*_{K_h h}\right) \in \Re^{Q \times K_h} \text{ with } q^*_{kh} = \frac{q_{kh}}{\theta_{kh}} - S^-_{qkh} \text{ and } k = 1 \ldots K_h$$
$$X_h{}^* = \left(x^*_{1h}, x^*_{2h}, \ldots, x^*_{K_h h}\right) \in \Re^{N \times K_h} \text{ with } x^*_{kh} = x_{kh} - S^-_{xkh} \text{ and } k = 1 \ldots K_h. \tag{12.13}$$

p_{kh}, q_{kh}, and x_{kh} being the original values.

II. *Stage Two*

A new frontier is estimated using the complete K-sized sample, but considering the following transformed (targets) data calculated in stage one:

$$\begin{aligned}
P^* &= \left(p^*_{1a}, p^*_{2a}, \ldots, p^*_{K_a a}, p^*_{1b}, p^*_{2b}, \ldots, p^*_{K_b b}\right) \in \Re^{PxK} \quad \text{with} \quad p^*_{kh} \in \Re^P \\
Q^* &= \left(q^*_{1a}, q^*_{2a}, \ldots, q^*_{K_a a}, q^*_{1b}, q^*_{2b}, \ldots, q^*_{K_b b}\right) \in \Re^{QxK} \quad \text{with} \quad q^*_{kh} \in \Re^Q \\
X^* &= \left(x^*_{1a}, x^*_{2a}, \ldots, x^*_{K_a a}, x^*_{1b}, x^*_{2b}, \ldots, x^*_{K_b b}\right) \in \Re^{NxK} \quad \text{with} \quad x^*_{kh} \in \Re^N
\end{aligned} \tag{12.14}$$

Therefore, the HD function for the environmental DEA models (12.11) and (12.12) is applied again with the aim of obtaining new scores for the whole sample, which are called

θ^*_k for $k = 1, \ldots, K$, and the slacks S^{*+}_{pk}, S^{*-}_{qk}, and S^{*-}_{xk} are also calculated as a result of the optimisation process.

These estimated values represent, for each firm k, the distance from its target (p^*, q^*, x^*) in its own frontier group (a or b) to the overall frontier. Note that different distances imply different productivities due to the non-discretionary variable. The new overall targets will be

$$\begin{aligned}
P^{**} &= \left(p^{**}_1, p^{**}_2, \ldots, p^{**}_K\right) \in \Re^{P\times K} \quad \text{with} \quad p^{**}_k = p^*_k \theta^*_k + S^{*+}_{pk} \quad \text{and} \quad k = 1 \ldots K \\
Q^{**} &= \left(q^{**}_1, q^{**}_2, \ldots, q^{**}_K\right) \in \Re^{Q\times K} \quad \text{with} \quad q^{**}_k = \frac{q^*_k}{\theta^*_k} - S^{*-}_{qk} \quad \text{and} \quad k = 1 \ldots K \\
X^{**} &= \left(x^{**}_1, x^{**}_2, \ldots, x^{**}_K\right) \in \Re^{N\times K} \quad \text{with} \quad x^{**}_k = x^*_k - S^{*-}_{xk} \quad \text{and} \quad k = 1 \ldots K
\end{aligned} \tag{12.15}$$

and the effect due to the non-discretionary variable z is calculated by means of the following expressions:

$$\begin{aligned}
\Delta P &= (\Delta p_1, \Delta p_2, \ldots, \Delta p_K) \in \Re^{P\times K} \quad \text{with} \quad \Delta p_k = p^{**}_k - p^*_k \\
\Delta Q &= (\Delta q_1, \Delta q_2, \ldots, \Delta q_K) \in \Re^{Q\times K} \quad \text{with} \quad \Delta q_k = q^{**}_k - q^*_k \\
\Delta X &= (\Delta x_1, \Delta x_2, \ldots, \Delta x_K) \in \Re^{N\times K} \quad \text{with} \quad \Delta x_k = x^{**}_k - x^*_k
\end{aligned} \tag{12.16}$$

III. *Stage Three*

To eliminate the effect of the non-discretionary variable, the original data are transformed by using the incremental values calculated in (12.16) in the following way:

$$\begin{aligned}
P^c &= \left(p^c_1, p^c_2, \ldots, p^c_K\right) \in \Re^{P\times K} \quad \text{with} \quad p^c_k = p_k + \Delta p_k \quad \text{and} \quad k = 1 \ldots K \\
Q^c &= \left(q^c_1, q^c_2, \ldots, q^c_K\right) \in \Re^{Q\times K} \quad \text{with} \quad q^c_k = q_k + \Delta q_k \quad \text{and} \quad k = 1 \ldots K \\
X^c &= \left(x^c_1, x^c_2, \ldots, x^c_K\right) \in \Re^{N\times K} \quad \text{with} \quad x^c_k = x_k + \Delta x_k \quad \text{and} \quad k = 1 \ldots K
\end{aligned} \tag{12.17}$$

Then, the HD function for the environmental DEA models (12.11) and (12.12) is applied again to the data (12.17) with the aim of obtaining the real efficiencies, having removed the non-discretionary variable effect. The results of the optimisation process are the following scores:

θ_k^c for $k = 1,\ldots,K$ and the slacks S_{pk}^{c+}, S_{qk}^{c-}, and S_{xk}^{c-}.

12.2.3.2 Quantifying the Non-discretionary Effect

As we have already seen in the development of the method, the effect due to the non-discretionary variable is related to the estimated distance in stage two (θ^*). We define this effect by

$$\hat{E}_k = \left(\frac{1}{\theta_k^*}\right) \times 100 \quad \text{for} \quad k = 1, \ldots, K \tag{12.18}$$

In order to evaluate the impact of the non-discretionary effect between the two subgroups, we calculate for each subgroup $h = (a, b)$ the average of this effect, and therefore,

$$\hat{E}_a = \frac{1}{K_a} \sum_{ka=1}^{K_a} \hat{E}_{ka}, \quad \forall DMU_{ka} \in \text{subsample } (h = a), \quad \text{and}$$

$$\hat{E}_b = \frac{1}{K_b} \sum_{kb=1}^{K_b} \hat{E}_{kb}, \quad \forall DMU_{kb} \in \text{subsample } (h = b) \tag{12.19}$$

And then we calculate the environmental ratio:

$$\text{ER} = \frac{\hat{E}_a}{\hat{E}_b} \tag{12.20}$$

The magnitude of this ratio indicates the importance of the non-discretionary variable in the efficiency assessment process.

12.3 Data and Efficiency Model

12.3.1 Source and Elaboration of Data

The primary source of data is the official record of the 806 oil mills in Andalusia in 2005–2006. These data contain about 30 variables. The most relevant ones for our study are those related to oil production, quantity of processed olives, extraction system, storage and treatment of effluents, and form of business organisation, among others. This information has been completed with two complementary sources that are necessary for the analyses: (1) the companies accounting reports in the register of companies and cooperatives and (2) a survey elaborated for a sample of companies like the one under study, so as to find out more about other issues related to socioeconomic, quality, and environmental impact aspects. After the sampling and the data cleaning processes, the records of 88 oil-mill industries are considered valid and complete, and they make up the final sample used in this study, 11% of the total census.

With the aim of studying the quality level and the environmental impact of the production process, each phase of the technological process (transport, reception and storage of the olive, extraction, storage of oil, and management of effluents) has been analysed. The variables that must be gathered from each company are chosen in order to get an overall value for each index, quality, and environmental impact. Nevertheless, as there are several aspects to be borne in mind, it is also necessary to define the priority relations among them, so that they can be included in the index with a specific level of relevance. This sequence has been established according to the opinions of 16 experts and after the application of a two-wave Delphi method (Almansa and Martinez-Paz 2011) that has improved the grade of consensus, as it reduces the dispersion of answers. Through a regular ranking in a Likert scale (0 null importance – 5 maximum importance), the experts assessed a wide range of attributes related to quality and environmental impact in the production process (Rikkonen 2005). This ranking determines the relevance and weight of every attribute in the index. The values for these attributes in each company have been compiled from the above-mentioned direct surveys. And finally, efficiency and quality indices have been constructed and presented in the Results section. These indices gradually increase on the aspect evaluated in the 0–100 interval.

12.3.2 The Technical Efficiency Model

The variables used in the efficiency model are shown in Table 12.1. The first two outputs, production (oil) and quality level in the production (quality index), are desirable products, while the environmental impact of the production process (Environmental Impact Index) is an undesirable product. The formulated model maximises desirable outputs and minimises the undesirable output, given the inputs of olive, labour, and capital (fixed and floating). The categorical variable *FOB* is an environmental, non-discretionary variable that divides the sample into two subsamples (cooperatives and corporate companies, respectively).

The oil production corresponds to the period under study (2005–2006). There was an attempt to break down the production by qualities, but the data provided by the companies were neither homogeneous nor comparable among them.

The fact that we use staff costs instead of the usual, physical employment variable (worked hours, number of full-time employees, etc.) is because, although the reports in registers are systematic and precise, the data referred to labour is not

Table 12.1 Variables of the efficiency model

Outputs	Oil (ton)
	Quality index
	Environmental Impact Index
Inputs	Grinded olive (ton)
	Floating capital (€)
	Fixed capital (€)
	Staff costs (€)
Environmental non-discretionary variable	Form of business organisations (*FOB*): cooperatives or corporate firms

Source: Prepared by the authors

Table 12.2 Descriptive statistics of outputs and inputs

	Mean	St. dev.	Minimum	Maximum
Grinded olive (ton)	5,530	5,080	254	21,482
Staff costs (10^3 €)	117	115	3	747
Floating capital (10^3 €)	3,336	2,940	177	12,011
Fixed capital (10^3 €)	116	95	3	487
Oil (ton)	1,644	1,330	185	5,229
Quality index	49.8	6.7	31.6	68.0
Environmental Impact Index	38.4	8.1	17.0	61.2

Source: Elaboration by the authors based on the surveys

homogeneous, and the reports use different methods to calculate this information, which makes this data unreliable and invalid for the model. In addition, there is no precise and systematic data on the quality of the labour factor. This lack has been made up by the incorporation of the second phase of the analysis with socioeconomic, non-discretionary variables that are related to the manager's training, the master's years of experience, and the temporary nature of the employment.

The floating capital has been measured by the total running and maintenance costs. This definition does not match the accounting concept of working capital (which is the current assets minus the current liabilities) but the operative concept of cash flow, meaning the consumable elements or goods in the production cycle of the firm. The fixed capital refers to the annual depreciation of the firm's fixed assets.

Prior to the efficiency analysis, a descriptive analysis of the involved variables has been carried out. Table 12.2 includes the description of oil production and the inputs of olive, labour, and capital with its two variables, floating and fixed capital.

In view of the results, a great variability in the magnitude of all variables can be observed. This is due to the fact that among the 88 oil mills under study, some of them are small and others are huge; therefore, the sample is balanced in the efficiency study according to this significant feature (Färe et al. 1994).

Table 12.3 shows a descriptive analysis of additional characteristics of the sample that will be considered in order to determine associative relations between the oil-mill profiles and the efficiency levels.

Table 12.3 Description of the socioeconomic non-discretionary variables

Continuous variables				
Variable	Mean	St. dev.	Minimum	Maximum
Age (years)	41.38	9.62	23	67
Master's seniority (years)	14.95	10.97	2	40
Proportion of permanent jobs (%)	61.21	21.45	15	100
Number of members (no.)	288	416	1	1,800
Dichotomous variables				
Variable	Yes (%)	No (%)		
Manager's specialised training	61.1	38.9		
Agricultural association membership	75.3	24.7		
Marketing association membership	36.0	64.0		
Internet sales	15.7	84.3		
Cooperative association	45.5	54.5		

Source: Elaboration by the authors based on the surveys

Table 12.4 Differences in the mean inputs and outputs regarding *FBO*

Variable	Cooperatives	Corporate firms	Sig. *t*-test
Grinded olive (ton)	7,354	4,010	0.00
Staff costs (10^3 €)	134	103	–
Floating capital (10^3 €)	4,648	2,243	0.00
Fixed capital (10^3 €)	154	84	0.00
Oil (ton)	2,181	1,197	0.00
Quality index	50.7	49.1	–
Environmental Impact Index	39.1	37.9	–
N	40	48	

Source: Elaboration by the authors based on the surveys

There are four variables measured continuously to be considered: the employer's age, the mill master's seniority, the ratio between fixed and temporary workers (%), and the number of members of the company. Some dichotomous variables are also studied: the existence of a manager specialised in the mill industry, membership to any agricultural or marketing associations, Internet sales, and the legal form (cooperative or others).

Table 12.4 includes the analysis of the differences in the mean variables between the two groups of firms that form the sample (cooperatives and corporate companies). This aims at testing the need for introducing the *FOB* variable as an environmental, non-discretionary variable.

There are significant differences in the means of olive grinded and oil produced and in the capital used, both floating and fixed. These means are higher in cooperatives. Ratios between outputs and inputs in each group may illustrate the great differences in apparent productivity. Then, it is of interest for the efficiency study to include the *FBO* variable, which has to do with the business structure. It can be concluded that the frontier (technology) of cooperatives may differ from that of corporations.

12.4 Results

The efficiency level of oil mills has been calculated by the resolution of the method proposed in the Methodology section. Despite this score has been defined as $\theta \geq 1$ in (12.11), we present the results of the efficiency index evaluated as $0 \leq (1/\theta) \leq 1$, in order to handle a more intuitive measure.

Table 12.5 shows the descriptive statistics of the efficiency indices deriving from this method application, under the assumptions of constant returns (CRS) and variable returns (VRS) in the last stage. Scale efficiency, which is the quotient between technical efficiency and pure efficiency, as well as the percentage of efficient firms for each measurement, is also included in Table 12.5.

The average efficiency levels are high, although the technical efficiency minimum is 0.65. There is a great percentage (27.3%) of completely efficient firms both technically and in scale. Given the specification of the model, the inefficiency level, evaluated as the hyperbolic distance at each firm's frontier, determines the possible improvements that could be carried out to increase production and quality and diminish the environmental impact.

As explained in the Methodology section, the measures taken from the solution of the method show the efficiency levels once the business structure effect is corrected. They actually measure the levels of oil mills, good or bad performances in the management of resources, and the distance to the frontier is not attributable to any of the differential factors that are characteristic to both business organisations.

In this way, it is possible to quantify the effect of the *FBO* variable by applying expression (12.19) to both groups. Table 12.6 shows the mean values ($\hat{E}_{\text{coo}}$ and $\hat{E}_{\text{cor}}$) together with the environmental ratio ER, which is obtained by expression (12.20). The value of the latter is 11.29, meaning that there exists a wide distance between the frontiers of both groups. The least productive group is the one that has a greater mean effect, that is, in this case, the group of cooperatives. Moreover, the difference in the means between the effects of both groups has been contrasted, and it is significant (P value = 0.001).

Table 12.5 Basic statistics of the efficiency indices

	Minimum	Maximum	Mean	St. dev.	Efficient firms (%)
Technical (CRS)	0.65	1.00	0.91	0.09	27.3
Pure (VRS)	0.66	1.00	0.95	0.07	51.1
Scale (SCA)	0.74	1.00	0.96	0.06	27.3

Source: Elaboration by the authors

Table 12.6 Effect of the FBO variable in efficiency

Group	FBO effect (%)	ER
Cooperatives	3.50	11.29
Corporate firms	0.31	

Source: Elaboration by the authors

Table 12.7 Differences in efficiency means between the two groups

	Cooperatives	Corporate firms	Sig. *t*-test
Technical (CRS)	0.82	0.89	0.00
Pure (VRS)	0.91	0.92	–
Scale (SCA)	0.90	0.97	0.00

Source: Elaboration by the authors

Table 12.8 Distribution of efficiency by groups

(No)	Cooperatives	Corporate firms	Total
Total	40	48	88
Constant returns (CRS)			
Efficient	2	17	19
1st quartile	23	24	47
Under 1st quartile	15	7	22
Variable returns (VRS)			
Efficient	2	17	19
1st quartile	23	24	47
Under 1st quartile	15	7	22
Scale returns (SCA)			
Efficient	6	30	36
1st quartile	19	14	33
Under 1st quartile	15	4	19

Source: Elaboration by the authors

Once the existence of different frontiers has been detected, the need to include the *FBO* variable is clear. Otherwise, if we had not done so, the efficiency estimation would have biased the results against cooperatives.

Tables 12.7 and 12.8 gather a set of contrasts found between the two groups in order to go deeper in the analysis of the differences in efficiency. In the first place, Table 12.7 shows the differences in the means of the three efficiency measures. It is noticeable that the mean efficiency is higher for corporate firms both in constant and scale returns, whereas there is no statistically significant difference (although in the same direction as the other two) in the model of variable returns.

These results imply that, even after correcting the *FBO* effect (which does not depend on the own management of the firm), cooperatives still are less efficient than the rest of the sample (constant returns), also in scale, which means that they are not at their optimum size. Thus, we can affirm that cooperatives are less efficient than corporate firms, which supports the prevailing hypothesis in the existing literature to this respect. However, it must be highlighted that the special treatment given in this study to business structures in the efficiency assessment allows corroborating this assertion even more strongly than in other works that compare efficiency levels among groups with no frontier separation.

To finish the comparison of the efficiency levels between the two groups, we have examined the amount of firms in each group that belong to three different categories:

Table 12.9 Tobit estimation of the efficiency factors

Variables	Coeff.	Sign.
Constant	0.905	0.00
No of members	−0.008	0.14
Master's seniority	−0.003	0.07
Manager's training	0.006	0.08
Internet sales	0.003	0.09
Agricultural associations membership	0.007	0.15
Marketing association membership	0.056	0.09
Proportion of permanent jobs	−0.005	0.08
McFadden's pseudo R^2	0.356	

Source: Elaboration by the authors

totally efficient firms, inefficient firms having an index higher that the first quartile, and inefficient firms with an index lower than the first quartile. Table 12.8 shows the information related to the classifications prepared for the indexes of constant, variable, and scale returns, respectively.

Among the 19 totally efficient firms within the CRS index, only two of them are cooperatives, while 17 are non-cooperative business. Actually, there are a much higher proportion of cooperatives in the category of the least efficient firms. When it comes to the model of low variable returns (VRS), although the overall differences are not significant, the proportion of the distribution goes in line with the CRS index results: among the efficient businesses, there are more trading companies than cooperatives; whereas, among the inefficient ones above the first quartile, there are less trading companies than cooperatives. As for scale efficiency, similar conclusions are drawn.

Therefore, thanks to these comparative analyses, we can assert that efficiency levels are lower in cooperative associations, even after having corrected the structural difference that prevent them from having more productivity. This means that there is room for improving the management of cooperatives.

And finally, a second-stage analysis has been carried out so as to study the possible associations between the efficiency index and the socioeconomic characteristics of every firm that were not included as inputs or outputs in the frontier formulation. This analysis is aimed at finding out the impact of these variables in the index of technical efficiency. Due to the limited nature both on the top and bottom of the efficiency index, the selected method is a doubly censored Tobit regression model, which is the alternative to avoid biased estimators related to the use of OLS regressions with this kind of data (Simar and Wilson 2007). The socioeconomic, non-discretionary variables included in the model (see Table 12.4) have been contrasted to detect multicollinearity, which would also bias parameters (Freese and Scott 2006). The estimation results of the efficiency level are included in Table 12.9.

In sight of these results, considering the statistical signification level at 10% and the sign of their coefficient, we conclude that the most efficient oil mills are those that have a young master of operations, a manager trained in business management, Internet sales, membership to marketing associations, and a low proportion of permanent jobs.

12.5 Conclusions

This work studies the level of technical efficiency in the Andalusian oil industry from a multi-output, non-parametric approach by conducting the data envelopment analysis (DEA) methodology with non-radial distance functions, as well as implementing environmental and non-discretionary variables.

The production frontier includes three outputs: quantity and quality of oil production, the outputs to be maximised, and one output to be minimised, the environmental impact of the production process. The inputs are the following: grinded olive, labour, and capital (both fixed and floating).

The analysis is carried out by including non-discretionary variables from two points of view. It is considered that the business structure (cooperative or corporation) of the firm affects the frontier (technology). This variable is included through a specific three-stage method. The relation between efficiency and other non-discretionary variables is analysed by the estimation of a Tobit model.

Having a sample of 88 oil-mill industries in Andalusia as the starting point, the indices for the two nonconventional outputs in this type of analysis are elaborated; quality is quantified by means of an aggregated index that gathers some aspects related to the separation of olives, critical points, and traceability. The environmental impact is assessed by another index that includes the effects produced on soil, water, air, and sound comfort.

Among the oil mills under study, there are two groups that differ from each other in their business structures: 42% are cooperatives and the rest of them are corporate trading companies. The descriptive study on inputs and outputs carried out before the efficiency analysis leads to the hypothesis that both groups might have a different frontier, and this has been corroborated by the results obtained.

The efficiency levels found, once the effect of the form of business organisation has been corrected, are high on average, around 90%. However, there are firms that could increase their quality and production levels up to 40% and shrink the environmental impact up to the same percentage, without enhancing their industrial facilities. As for the scale efficiency, just 27% of firms are working at their optimum size. Nevertheless, scale inefficiency is not really high in oil-mill industries that do not work with constant returns.

We hold that the effect of the business structure is significant, which justifies the use of the suggested method. Regardless of the correction in the effect of the different frontier, cooperatives are less efficient than corporate companies. The problem of scale inefficiency presents the differences and then the chances for improvement plans in the clearest way. In this respect, Oustapassidis et al. (1998) hold that low scale efficiency in Greek dairy cooperatives is due to the fact that an excess of inputs is better accepted by cooperatives than by corporate companies because inputs come from their owners.

According to the results in the last stage and the existence of scale inefficiencies, collaboration agreements between firms are highly advisable. For instance, the externalisation of some processes like the product marketing and advertising is a model that deserves some attention in order to improve the sector's efficiency. This activity would create entities offering services to several companies and would facilitate the inclusion of two efficiency factors: on-line sales and business management experts. The reason why the inclusion of masters with less seniority is also a symptom of more efficiency could be the fact that training and updated knowledge are more useful than experience in techniques and processes when it comes to managing resources in an efficient way.

The fact that the labour temporality is a factor that enhances efficiency could be striking. But flexibility in the number of employees is crucial when planning an efficient allocation of resources, owing to the special features of this extraction industry, such as the important seasonal component.

There is no doubt that the special idiosyncrasy of cooperative entities, where employer and provider are identical, determines the processes of resource management and the lower technical efficiency. On the one hand, cooperatives put people, not capital, in the core of the business, meaning that the decisions made by cooperatives are meant to balance the profitability objects and the interests of their members and sometimes even of the community where they are located in (Soboh et al. 2009). On the other hand, there is no doubt that the extra services that many of these entities provide (capital financing, accounting management, etc.) could be regarded as another output or, at least, as a positive externality of the production process in future research.

References

Abdourahmane T, Bravo-Ureta B, Rivas T (2001) Technical efficiency in developing country agriculture: a meta-analysis. Agric Econ 25:235–243

Ali AI, Seiford LM (1993) The mathematical programming approach to efficiency analysis. In: Fried HO, Lovell CAK, Schmidt SS (eds) The measurement of productive efficiency: techniques and applications. Oxford University Press, Oxford

Almansa C, Martinez-Paz JM (2011) Intergenerational equity and dual discounting. Environ Dev Econ 16(6):685–707

Alonso M, García J (2009) Cooperatives versus corporate in the Spanish agricultural sector: non-parametric estimation of technical efficiency. Anales de estudios económicos y empresariales 19:61–90

Álvarez A (ed) (2001) La medición de la eficiencia y la productividad. Pirámide, Madrid

Amores A, Contreras I (2009) New approach for the assignment of new European agricultural subsidies using scores from data envelopment analysis: application to olive-growing farms in Andalusia (Spain). Eur J Oper Res 193(3):718–729

Ball VE, Lovell CAK, Nehring R, Somwaru A (1994) Incorporating undesirable outputs into models of production: an application to US agriculture. Cahiers d'Economie et Sociologie Rurales 31:60–74

Ball VE, Lovell CAK, Luu H, Nehring R (2004) Incorporating environmental impacts in the measurement of agricultural productivity growth. J Agric Resour Econ 29:436–460

Banker RD, Morey RC (1986) Efficiency analysis for exogenously fixed input and outputs. Oper Res 34(4):513–521

Barnes A (2006) Does multi-functionality affect technical efficiency? A non-parametric analysis of the Scottish dairy farms. J Environ Manage 80(3):287–294
Barreiro J, Martínez-Paz J, Espinosa-Goded M (2009) Impact of agri-environmental scheme participation on dry-land cereal farms' efficiency. ITEA: Información Técnica Económica Agraria 105(3):184–201
Bartlett W, Cable J, Estrin S, Jones DC, Smith SC (1992) Labor-managed cooperatives and private firms in North Central Italy: an empirical comparison. Ind Labor Relat Rev 46(1):103–118
Battese E (1992) Frontier production functions and technical efficiency: a survey of empirical applications in agricultural economics. Agric Econ 7:185–208
Baumgärtner S (2004) Price ambivalence of secondary resources: joint production, limits to substitution, and costly disposal. Resour Conserv Recycl 43:95–117
Bonin JP, Jones DC, Puttermann L (1993) Theoretical and empirical studies of producer cooperatives: will ever the twain meet? J Econ Lit 31(4):1290–1320
Brada J, Méndez J (2009) Technology, effort and the efficiency of production: labor-managed versus capitalist firms. South-East Eur J Econ 1:35–53
Bravo-Ureta B, Pinheiro A (1993) Efficiency analysis of developing country agriculture: a review of the frontier function literature. Agric Resour Econ Rev 22(1):88–101
Bravo-Ureta B, Solís D, Moreira-López V, Maripani J, Thiam A, Rivas T (2007) Technical efficiency in farming: a meta-regression analysis. J Prod Anal 27(1):57–72
Charnes A, Cooper W, Rhodes E (1981) Evaluating program and managerial efficiency: an application of data envelopment analysis to program follow through. Manage Sci 27(6): 668–697
Cook ML (1995) The future of agricultural cooperatives: a neo-institutional approach. Am J Agric Econ 77(5):1153–1159
Cooper WW, Seiford LM, Zhu J (2004) Handbook on data envelopment analysis. Springer, Boston
Damas E, Romero C (1997) Análisis no paramétrico de la eficiencia relativa de las almazaras cooperativas en la Provincia de Jaén durante el período 1975–1993. Revista Española de Economía Agraría 180:279–304
Daraio C, Simar L (2005) Introducing environmental variables in nonparametric frontier models: a probabilistic approach. J Prod Anal 24(1):93–121
Dios-Palomares R, Martínez-Paz JM (2011) Technical, quality and environmental efficiency of the olive oil industry. Food Policy 36(4):526–534
Dios-Palomares R, Martínez-Paz JM, De Haro-Jiménez T (2005) Accounting for technical, quality and environmental efficiency in the olive oil industry. Paper presented at the ninth European workshop on efficiency and productivity analysis, Brussels, Belgium
Dios-Palomares R, Martínez-Paz JM, Martínez-Carrasco F (2006) El análisis de eficiencia con variables de entorno: un método de programas con tres etapas. Estudios de Economía Aplicada 24:1–23
Faber M, Props J, Baumgärtner S (1998) All production is joint production –a thermodynamic analysis. In: Faucheax S, Gowdy J, Nicolaï I (eds) Sustainable and firms: technological change and the changing regulatory environment. Edwars Elgar, Cheltenham, pp 131–158
Färe R, Grosskopf S (2004) Modeling undesirable factors in efficiency evaluation: comment, European. J Oper Res 157:242–245
Färe R, Grosskopf S, Lovell CAK (1985) The measurement of efficiency of production. Kluwer-Nijhoff, Boston
Färe R, Grosskopf S, Lovell CAK, Pasurka C (1989) Multilateral productivity comparisons when some outputs are undesirable: a non-parametric approach. Rev Econ Stat 75:90–98
Färe R, Grosskopf S, Lovell C (1994) Production frontiers. Cambridge University Press, Cambridge
Färe R, Grosskopf S, Noh D, Weber W (2005) Characteristics of a polluting technology: theory and practice. J Econom 126:469–492
Ferrier GD, Porter PK (1991) The productive efficiency of US milk processing cooperatives. J Agric Econ 42:161–173

Freese J, Scott J (2006) Regression models for categorical dependent variables using stata. Stata Press, College Station
Fried H, Schmidt SS, Yaisawarng S (1999) Incorporating the operating environment into a nonparametric measure of technical efficiency. J Prod Anal 12(3):249–267
Fried H, Lovell C, Schmidt SS, Yaisawarng S (2002) Accounting for environmental effects and statistical noise in data envelopment analysis. J Prod Anal 17:157–174
Gázquez J, Sánchez M (2009) Factors influencing olive oil brand choice in Spain: an empirical analysis using scanner data. Agribusiness 25(1):36–55
Giannakas K, Tran K, Tzouvelekas V (2000) Efficiency, technological change and output growth in Greek olive growing farms: a Box-Cox approach. Appl Econ 32:909–916
Hailu H, Veeman V (2001) Non-parametric productivity analysis with undesirable outputs: an application to the Canadian pulp and paper industry. Am J Agric Econ 83:605–616
Hart O, Moore J (1990) Property rights and nature of the firm. J Polit Econ 98(6):1119–1158
Hernández-Sancho F, Picazo-Tadeo A, Reig-Martinez E (2000) Efficiency and environmental regulation. Environ Resour Econ 15:365–378
Hofler RA, Payne JE (1993) Efficiency in social versus private agricultural production: the case of Yugoslavia. Rev Econ Stat 73:153–157
Jensen MC, Meckling WH (1979) Rights and production function: an application to labor-managed firms and codetermination. J Bus 52(4):469–506
Jones D (2007) The productive efficiency of Italian producer cooperatives: evidence from conventional and cooperative firms, in Sonja Novkovic. In: Sena V (ed) Cooperative firms in global markets, vol 10, Advances in the economic analysis of participatory & labor-managed firms. Emerald Group Publishing Limited, London, pp 3–28
Karagiannis G, Tzouvelekas V (2009) Measuring technical efficiency in the stochastic varying coefficient frontier model. Agric Econ 40:389–396
Karagiannis G, Kien C, Tzouvelekas V (2003) On the choice of functional form in stochastic frontier modelling. Empir Econ 28(1):75–100
Lambarraa F, Serra T, Gil J (2007) Technical efficiency analysis and decomposition of productivity growth of Spanish olive farms. Span J Agric Res 5(3):259–270
Lambarraa F, Stefanou S, Gil J (2009) The analysis of long-run technical inefficiency in Spanish olive Sector. Paper presented at the IV Congreso de Eficiencia y Productividad, Córdoba, Spain
Lovell CAK, Pastor J (1995) Units invariant and translation invariant DEA models. Oper Res Lett 18:147–151
Lozano-Vivas A, Pastor JT, Pastor JM (2002) An efficiency comparison of European banking systems operating under different environmental conditions. J Prod Anal 18:59–77
Maietta O, Sena V (2008) Is competition really bad news for cooperatives? Some empirical evidence for Italian producers' cooperatives. J Prod Anal 29(3):221–233
Mesias FJ, Martinez-Carrasco F, Martinez-Paz JM, Gaspar P (2011) Functional and organic eggs as an alternative to conventional production: a conjoint analysis of consumers' preferences. J Sci Food Agric 91(3):532–538
Mili S (2009) Market dynamics and policy reforms in the olive oil sector: a European perspective. In: Noronha MT, Nijkamp P, Rastoin JL (eds) Traditional food production facing sustainability: a European challenge. Ashgate Publishing, Hampshire
Millán J (1986) Eficiencia, dimensión y crecimiento de las cooperativas olivareras de Jaén. Ph. dissertation, E.T.S.I.A.M, Universidad de Córdoba
Muñiz MA (2001) ¿Son realmente menos eficientes los centros LOGSE? (La evaluación DEA de los Institutos de Educación Secundaria). Hacienda Pública Española 157:169–196
Oustapassidis K, Vlachvei A, Karantininis K (1998) Growth of investor owned and cooperatives firms in Greek dairy industry. Ann Public Co-op Econ 69:399–417
Picazo-Tadeo A, Prior D (2009) Environmental externalities and efficiency measurement. J Environ Manage 90(11):3332–3339
Piesse J, Thirtle C, Turk J (1996) Efficiency and ownership in Slovene dairying: a comparison of econometric and programming technique. J Comp Econ 22(1):1–22

Prieto A, Zofío JL (2004) Modelización de la gestión ambiental preventiva mediante estándares. Investigaciones económicas 28:43–66
Rallo L (2010) El olivar en un tiempo de cambio. Citoliva, Sevilla
Rikkonen P (2005) Scenarios for future agriculture in Finland: a Delphi study among agri-food sector stakeholders. Agric Food Sci 14(3):205–223
Salazar T, Galve C (2008) Cooperativa vs. Capitalista: Afecta la forma de gobierno a la eficiencia productiva? Revista Europea de Dirección y Economía de la Empresa 17:133–144
Sánchez J, Gallego V, Araque E (2011) El olivar andaluz y sus transformaciones recientes. Estudios Geográficos 270:203–229
Schmitt G (1993) Why is the agriculture of advanced western countries still organized by family farms? Will this continue to be so in the future? Eur Rev Agric Econ 18(3):443–458
Seiford LM, Zhu J (2002) Modeling undesirable factors in efficiency evaluation. Eur J Oper Res 142:16–20
Simar L, Wilson P (2007) Estimation and inference in two-stage, semi-parametric models of production processes. J Econom 136(11):31–64
Soboh R, Oude A, Giesen G, Van Dijk G (2009) Performance measurement of the agricultural marketing cooperative: the gap between theory and practice. Rev Agric Econ 31:446–469
Tzouvelekas V, Pantzios C, Fotopoulos C (2001) Technical efficiency of alternative farming systems: the case of Greek organic and conventional olive-growing farms. Food Policy 26(6):549–569
Zaim O (2004) Measuring environmental performance of state manufacturing through changes in pollution intensities: a DEA framework. Ecol Econ 48:37–47
Zaim O, Taskin F (2000) A Kuznets curve in environmental efficiency: an application on OECD countries. Environ Resour Econ 17:21–36
Zhou P, Poh KL, Ang B (2007) A non-radial DEA approach to measuring environmental performance. Eur J Oper Res 178:1–9
Zhu J (2009) Modeling undesirable measures. In: Zhu J (ed) Quantitative models for performance evaluation and benchmarking. Springer, New York, pp 97–107
Zofío JL, Prieto A (2001) Environmental efficiency and regulatory standards: the case of CO2 emissions from OECD industries. Resour Energy Econ 23:63–86

Chapter 13
Efficiency Assessment: Final Remarks

Jorge Santos, Emiliana Silva, and Armando B. Mendes

Abstract In this concluding chapter the editors make some important remarks on the importance and conclusions of the book as a whole, focusing on its importance, relevance and adequacy to an extended audience of researchers in the efficiency analysis in the agricultural and environment fields.

Keywords Data envelopment analysis • Efficiency analysis • Agriculture • Environment

This book introduces the techniques most applied in the assessment of efficiency in the agricultural and environmental field. It includes the basic concepts of efficiency analysis but goes deep in some recent and innovative techniques, both on the non-parametric methods and in the parametric ones. It is pointed out that data envelopment analysis, originally applied to ex post analysis of data, can also be applied on ex ante analysis, being a serious competitor against traditional methods like multi-criteria decision making and cost benefit analysis. In the stochastic

J. Santos (✉)
CIMA, Departamento de Matemática Colégio Luís Antº Verney R., Universidade de Évora, Romão Ramalho, 59 7000-671 Évora, Portugal
e-mail: jmas@uevora.pt

E. Silva
CEEAplA, Departamento de Ciências Agrárias, Universidade dos Açores, Rua Capitão João de Ávila, 9700-042 Angra do Heroísmo, Azores, Portugal
e-mail: emiliana@uac.pt

A.B. Mendes
CEEAplA, Departamento de Matemática, Universidade dos Açores, Rua da Mãe de Deus, 9500-801 Ponta Delgada, Azores, Portugal
e-mail: amendes@uac.pt

A.B. Mendes et al. (eds.), *Efficiency Measures in the Agricultural Sector: With Applications*, DOI 10.1007/978-94-007-5739-4_13,

frontier analysis chapter, a very concise and exhaustive presentation of the several models is made. An extra chapter about superefficiency and weight adjustment ends the introductory section.

Concerning the applications, their main focus is the Iberian Peninsula and also with greater emphasis the Azorean archipelago.

The applications range from the analysis of agricultural grazing systems in Azores to olive oil production industry in Andalusia. This book aggregated the efforts of researchers of a dozen different universities from three different European countries.

A major concern with this book was to disseminate the use of free software, and in fact for the data envelopment analysis, both the EMS from Holger Scheel and the DEAP from Coelli were used, but mainly the open source libraries of R. For the stochastic frontier analysis, the computer program FRONTIER Version 4.1 developed by Tim Coelli was the main choice. For the statistical analysis, since it was not the main goal of this book, some commercial software was used, but R was widely used too.

Efficiency in agriculture is a very important matter, for some reason the first DEA application by Farrell was with United States farms' data in 1957.

Agriculture is a major sector especially nowadays with a terrible demographic pressure that our planet is facing. But, agriculture is much more than just food production. With increasing global demand for transport energy, just a few years ago, we had to deal with a rise on food prices, because agriculture is also producing raw material for biofuels.

In fact, the situation was described synthetically by Tilman in the following statement:

> food, energy and environment trilemma. Tilman et al. (2009)

In fact all these three factors compete for the same resource: land. In Europe since 1950, some countries even adopted a Common Agricultural Policy (CAP), with huge importance in their budgets. The Common Agricultural Policy has two pillars: the first pillar is related with market intervention, and the second with the structural policy; this second pillar is increasing its importance specially after the MacSharry reform, where the productivist CAP is giving place to the fight against climate change; supporting employment, growth and food security; and promoting the environmental protection and rural development plans for avoiding desertification.

One of the main messages at the XIII World Congress of Rural Sociology (The New Rural World: From Crises to Opportunities) held in Lisbon (Portugal) during July 29 to August 4, 2012, was:

> Food comes first than agriculture.

This reinforces the previous trilemma; in fact in this overpopulated planet, hunger and malnutrition are serious problems, and the final goal is to eliminate world hunger. The supply probably is enough for the actual demand of food, but the distribution is not adequate. Efficiency is the way to promote the increasing agricultural production and to provide food to everyone.

It is important to notice that, exactly the same way, economic crisis settled down in developed countries; it would not be a big surprise if scarcity of food happened too in countries with unbalanced age trees and dependent on external sources of energy.

Anyway, demographic measures should be taken so that according to 'land capacity' (it should be considered on a more global sense, as economic capacity), we should have a sustainable birth rate. In fact, the poorest countries are those with higher birth rates, especially in Africa, and this leads to rather unbalanced distribution of food and wealth.

It is our hope that this book can contribute to the theory of efficiency analysis but especially to real problem application.

Reference

Tilman D, Socolow R, Foley JA, Hill J, Larson E, Lynd L, Pacala S, Reilly J, Searchinger T, Somerville C, Williams R (2009) Beneficial biofuels-the food, energy, and environment trilemma. Science 325(5938):270–271. doi: 10.1126/science.1177970

Index

A.B. Mendes et al. (eds.), *Efficiency Measures in the Agricultural Sector: With Applications*, DOI 10.1007/978-94-007-5739-4,

MIX
Papier aus verantwortungsvollen Quellen
Paper from responsible sources
FSC® C105338

If you have any concerns about our products,
you can contact us on
ProductSafety@springernature.com

In case Publisher is established outside the EU,
the EU authorized representative is:
Springer Nature Customer Service Center GmbH
Europaplatz 3, 69115 Heidelberg, Germany

Printed by Libri Plureos GmbH
in Hamburg, Germany